D0450322

THE CLIMATE NEXUS

THE CLIMATE NEXUS

Water, Food, Energy and Climate in a Changing World

Dr. Jon O'Riordan

Robert William Sandford

First Edition

RMB | Rocky Mountain Books Ltd.
rmbooks.com
@rmbooks
facebook.com/rmbooks

Cataloguing data available from Library and Archives Canada

ISBN 978-1-77160-142-9 (bound)
ISBN 978-1-77160-143-6 (epub)

Printed and bound in Canada

Distributed in Canada by Heritage Group Distribution and in the U.S. by Publishers Group West

For information on purchasing bulk quantities of this book, or to obtain media excerpts or invite the author to speak at an event, please visit rmbooks.com and select the "Contact Us" tab.

RMB | Rocky Mountain Books is dedicated to the environment and committed to reducing the destruction of old-growth forests. Our books are produced with respect for the future and consideration for the past.

We acknowledge the financial support of the Government of Canada through the Canada Book Fund and the Canada Council for the Arts, and of the province of British Columbia through the British Columbia Arts Council and the Book Publishing Tax Credit.

Nous reconnaissons l'aide financière du gouvernement du Canada par l'entremise du Fonds du livre du Canada et le Conseil des arts du Canada, et de la province de la Colombie-Britannique par le Conseil des arts de la Colombie-Britannique et le Crédit d'impôt pour l'édition de livres.

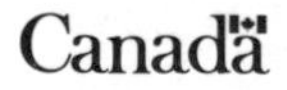

An initial step in the long journey of
educating how we can understand and
thus adapt to changes in the nexus.

— DR. JON O'RIORDAN

For Cara Nelson and all the 20- and
30-somethings out there who will inherit the
nexus and be the next generation's leaders
in realizing a vision of a better world.

— ROBERT SANDFORD

CONTENTS

ACKNOWLEDGEMENTS

The lead authors wish to acknowledge the support of the advisory board for Simon Fraser University's Adaptation to Climate Change Team (ACT), in particular Professor Nancy Olewiler in the School of Public Policy; Dr. Stewart Cohen, senior researcher with Environment Canada and adjunct professor of Forest Resources Management at the University of British Columbia; and Doug McArthur, director of the School of Public Policy at Simon Fraser. These experts have offered wise advice, astute scientific perspectives and valuable direction to all who have been involved in our work on the changing adaptation.

The authors also wish to acknowledge that this book would not have come into existence without the tireless support of ACT Executive Director Deborah Harford.

We are also grateful for the hard work and dedication of researchers Chelsea Notte, Sukhraj Sihota and Anthony Cotter of the School of Public Policy at Simon Fraser University, who contributed enormously to the content and perspectives put forward in this book.

The authors further wish to acknowledge and thank the Real Estate Foundation of British Columbia for its support of ACT's work on climate change adaptation and water governance, without which this book would not have been possible; and the Pacific Institute for Climate Solutions, whose support of Sukhraj Sihota's work was invaluable.

The authors accept sole responsibility for any errors, omissions, misperceptions or misunderstandings.

INTRODUCTION

The Age of Consequences

We cannot continue living as we do now. Why? Because we are dismantling our planet's biodiversity-based life-support system – the system that makes life possible on this planet. Through our numbers and activities we have lit a biogeochemical fuse that may cause the Earth system to move to a new and different state. While it has been easy to believe that the effects and consequences of biodiversity-based planetary life-support function are affecting some-one else, somewhere else, they are in fact happening to all of us – now. The focus of this book is about nexus – the interconnecting point where water, food, energy and climate become one; where what we do to and with any one of these factors affects the others. This nexus lies at the very heart of current civilization; it is ground zero in the fight on climate and hydrological change. This book, however, is also about another nexus: the place where the present and the future meet, the historical node at which humanity has an opportunity to work toward what the world might be like at its future and ultimate best.

While the Earth system will adjust, it may do so in a way that is hostile even to human existence. If we keep moving in the direction we are going we may find ourselves in a world in which we cannot survive without rapid adaptation and the evolution of technologies to support us. In such a world, we would have to reproduce artificially the services nature currently provides for free, with the aim of recreating the conditions in which we evolved before climate disruption and mass extinctions affect the Earth system and bring about the collapse of ordered human existence on the planet.

Careful examination of how rapidly the Earth system is changing shows that we will have to reframe our situation so that we can adapt quickly and effectively. One way in which we are shifting perspective is through the realization that we have entered a new geological era in which human activities rival the larger processes of nature. This new geological era is being called the Anthropocene. It is also being called the Age of Consequences. Unlike previous epochs – heralded by meteorite strikes and other geological events that resulted in mass extinctions – this epoch is marked by human impacts on the Earth system. Climate disruption is only one of these impacts. By virtue of our numbers and our activities, we have altered global carbon, nitrogen and phosphorus cycles. We are causing changes in the chemistry, salinity and temperature of the oceans and the composition of the atmosphere. These changes, in tandem with land-use changes and our growing water demands, have also altered the global water cycle. The cumulative measure of the extent to which we have changed these systems is the rate of biodiversity loss.

We have entered an era in which we can no longer count on heretofore natural processes to absorb human impacts on Earth-system function. We have to assume responsibility for directing previously self-regulating functions of the Earth system. Since we have disrupted the stability of our biodiversity-based planetary life-support system, we are going to have to recreate those functions ourselves. While we can briefly create conditions suitable for life in a space station, we do not yet know how to do this on Earth. If we continue on the path we are on now, there will not be enough water, food or energy, let alone adequate biodiversity, to sustain prosperity as we know it today, Whether or not we want to continue in the direction of the Anthropocene is something we urgently need to discuss as a global community.

There are two ways we can react to this emergency. We can continue to ignore the problem and fabricate the next human age out of the pieces of an ecologically diminished and climatically altered world; or humanity can choose to follow a new narrative, one that leads to creating a new and better world for us and for those who will come after us. This new narrative is called "adaptation" in this book. Humanity is at once the most potentially destructive of non-natural Earth-altering forces and yet the most adaptive and ingenious species to inhabit the planet. Though it is equally relevant to all of North America, and indeed to other parts of the globe, this book focuses on the ways in which we might advance sustainability in western Canada.

We are already past the point where changes in the carbon concentrations in the atmosphere will have profound and lasting effects on atmosphere and oceans that will forever change the way water, food, energy and climate interact. Adaptation will thus become a central plank in global policy-making, requiring a profound change to the way we currently do business over the coming decades. The book charts how this new focus on adaptation will transform how we use the resources in the nexus and how policy, and political, economic and social systems will have to adjust.

It should be no surprise that we find water at the heart of the nexus between food and energy and between the present and the future. Evaporation from the global ocean drives the hydrological cycle that supplies the fresh water that all terrestrial life forms require for their existence. Water makes life possible, and life in turn shapes the nature and character of Earth's biodiversity systems.

One of the great breakthroughs in the Earth sciences in the past century is the realization that what we thought were mere physical or cosmological processes, such as the evolution of Earth's atmosphere, are in part the creation of living things over time. In elementary school we learn that the planet's atmosphere is composed of 78 per cent nitrogen, 21 per cent oxygen and 1 per cent argon, with a few traces of carbon dioxide, neon, helium, methane, krypton, hydrogen, nitrous oxide, xenon, ozone, iodine, carbon monoxide, and ammonia. From this description, one might surmise that the atmosphere is nothing more than a transparent swirl of inert gases that somehow came into equilibrium through physical interaction of elements present at the birth of the planet. When we look up into the blue of the sky, we see not just air but the suspended residue of every geological event and the cumulative exhalation of every life form and ecological process that has ever taken place on or near Earth's surface and oceans since the beginning of time. As Charles David Keeling's curve – the graph of carbon dioxide concentrations in our atmosphere over the past half-century – so elegantly demonstrates, our atmosphere is, in part, the exhaled breath of life on Earth – a breath made possible by water.

We have also discovered that the larger Earth system of which the atmosphere is but a part is equally sensitive and fragile. Four components in tandem appear to regulate the stability that makes life possible here: the atmosphere; the hydrosphere, the oceans and the water we drink; the cryosphere, the climate-stabilizing refrigerating influence of Earth's snow, glaciers and ice sheets; and the biosphere, Earth's living elements. It is within the conditions created by these four fundamental Earth-system processes that all of humanity relentlessly expends its energies securing the water, growing the food and powering the economies upon which our current civilization depends for social and political stability. We cannot do without a proper functioning nexus if we are to enjoy anything resembling our current prosperity in the future and one day realize the dream of sustainability.

This book tells the story of the coming crisis at the convergence of water, food and energy caused by global changes in both climate and Earth's natural systems. The nexus is the stuff of all communities; if it fails, communities as we know them will also fail. It also charts a course of adaptation mechanisms that will be required in order to adjust how we use resources, such that this vital nexus can become sustainable even in the face of the profound changes to Earth's hydrologic and natural systems.

The thinking presented here is based partly on a synthesis of the five major reports prepared by Simon Fraser University's Adaptation to Climate Change Team (ACT) since 2006 on climate change adaptation and biodiversity and ecosystem services, extreme weather events, energy, food and crop security and water governance. Each of these reports analyzes the effects of a changing climate on the elements of the nexus separately, laying down a strong foundation of understanding about the challenges facing each one and proposing thoughtful suggestions for those charged with governance of these issues. It is upon this foundation that we now bring together our conclusions regarding the risks to the nexus in all its complexity, with the aim of providing a current picture of the emerging trade-offs between the elements and proposals for significant transformations in policy and practice that can help human society adapt.

The geographic scope of the analysis in this book is partly focused on British Columbia, where ACT is based and where a significant portion of its research has been undertaken. It is also focused on other regions in Canada in which its members have been particularly active with respect to water and water-related climate issues that connect with food production, energy generation and use, and issues related to biodiversity loss and Earth-system decline. To a lesser extent, this book includes examples from Canada as a whole, as there has been some progress – federally and provincially – on both mitigation and adaptation policies. The third plane of reference is North America, due to the importance of US policy and actions in this regard, especially under the Obama administration.

Presenting an economically developed nation such as Canada as an example here has the advantage of allowing the authors to demonstrate the current progress in reducing carbon emissions and tackling responses to the changing climate where there are strong governance institutions, available financial and human resources, and developed technology and professional service sectors. In short, in British Columbia and Canada, one would expect progress and focus in responding to the emerging tensions in the nexus. But as we will demonstrate in the coming chapters, there are many barriers to the kind of progress that is required for dealing purposefully with the impending crisis in the nexus. The authors explore the actions needed in Canada and, to some extent, the much greater challenges that face less developed economies.

The book is structured around five themes. The first chapter outlines why there is growing tension at the place where all these elements meet and interact. Chapter two elucidates the repeated warnings we have received about how climate disruption and loss of biodiversity undermine the productivity and viability of the nexus. The third chapter explains how changes in the nexus affect Canadians and what we are doing at present to address these challenges.

Chapter four confronts the costs and consequences for our economy and way of life should we fail to meet these challenges. Finally, chapter five attempts to describe the ultimate shape of the nexus: the world we want the next generation to inherit. The book concludes with a blueprint for action.

It is an understatement to say that this book was written with a sense of urgency. Early this year NASA reported that the period from January to April of 2015 was the hottest on record, a finding that was confirmed a few days later by the National Oceanic and Atmospheric Administration (NOAA). Subsequently, NOAA updated its "Global Ocean Heat and Salt Content" data webpage, observing, among several other parameters, that the oceans' heat content down to 2000 metres (1.24 miles) has been soaring – it had nearly gone off the chart during this writing.

There is a very real fear among experts that we have blown the biogeochemical fuse that controls planetary land and sea-surface temperatures and that we are now passing over what to us was an invisible threshold into a new global hydro-climatic state. In other words, climate change may have already gotten away on us. We are all in this together; and there is no place to hide. If there was ever a point in human history where it has been critical to take stock of exactly where we stand with respect to the interconnected realities of water, food, energy and climate in our planetary life-support system function – it is now.

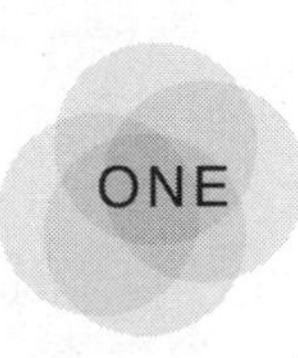

Nervousness at the Nexus: Tensions at the Centre Where All Things Meet

In his book *The Age of Consequences: A Chronicle of Concern and Hope*, Courtney White compares the Anthropocene to a hurricane building up slowly over warming water that is approaching the shore. We don't know exactly when or where the hurricane will make landfall, but when it does it will be destructive. White urges us to lower the wind speed of this gathering storm by increasing human resilience to the inevitability of its impacts. This, White argues, can be achieved by reversing ecosystem decline, creating sustainable prosperity and localizing sources of food and energy production. In short, White describes a nexus: a connection or series of connections linking two or more things. A nexus is also a hub or the centre of a series of connections that affect a range of other factors.

The nexus of water, food and energy is the focus of this book because these three resources are fundamental building blocks for all global economies and communities. Each of them is under increasing pressure, even without a changing climate, due to relentless population growth; removal of forest and soil cover due to land development; increasing rates of per capita consumption that are projected to continue rising over the next several decades; and accelerating changes we have made to the composition and behaviour of Earth's atmosphere. This nervousness at the nexus is further compounded by loss of natural biodiversity through widespread extinction, which is presently occurring a thousand times faster than average, considering what we know about Earth's long history. The grim reapers of biodiversity loss are also causing tensions elsewhere at the nexus: habitat destruction, invasive species, pollution, population growth and overharvesting and resource extraction.

Loss of biodiversity means a decline in the quality and range of ecosystem services that nature presently provides for free, not only to humans but to all life on Earth. For example, healthy soil recycles dead matter into new life; purifies wastes; circulates nutrients; captures, stores and purifies water; and captures and sequesters huge volumes of atmospheric carbon. Wetlands filter contaminated water and make it available for other uses; they buffer runoff from intense rainstorms, thus reducing the economic cost of flooding; and they store large quantities of carbon that would otherwise be released into the atmosphere. Forests and tree cover provide a similar range of ecosystem services: prevention of soil erosion, storage of nutrients and carbon, and provision of shade to protect against extreme heat. Shade is especially important in cities, where "heat island" effects are accelerating and ever-increasing amounts of energy are required for air conditioning. All ecosystems protect the species with which we share Earth and on which we depend for food and other services, quite apart from their own right to exist. Unfortunately, human activity has severely eroded the healthy functioning of all of these ecosystem services. One-third of the world's potentially farmable soils have been lost to erosion, much of it in the past 40 years. Vast areas of forest have been transformed into agricultural land. Wetlands have been drained to increase food production; vegetation has been paved over in urban areas, resulting in flash flooding or depletion of groundwater recharge as runoff is piped directly into rivers and oceans. The cumulative effect of the diminishment of these natural ecosystem services is a decline in the ability of the planet's biodiversity-based life-support system to remain stable and functioning predictably in the face of growing human interference with long-established natural processes.

Loss of natural biodiversity has another profound effect on Earth's natural systems: a reduced capacity to store carbon. Even as human activity increases carbon dioxide emissions, we decrease Earth's ability to store the released carbon. We will have to focus many adaptation policies on restoring and enhancing Earth's capacity to store carbon.

The links between the elements of the nexus are clear. Water is essential for food production and is a key component in energy production, from traditional fossil fuels (for example, oil and gas extraction and processing) to renewables (hydropower, solar, wind, and biofuels). Energy is needed for food production and distribution and to power harvesting, transportation, packaging and marketing. Water and wastewater treatment, water distribution and irrigation and new technologies like desalinization all require energy, some of which is now supplied by food crops, with food affordability ramifications for low-income people and intensified water pollution from agricultural chemicals. In addition, the elements of the nexus combine synergistically; their interactions with a changing climate are not simply additive. They compound one another in non-linear or logarithmic ways that result in magnification of their combined effects. Much of the analysis to date has been on the individual elements of the nexus. It is only recently that we have recognized the linkages among these elements, and the crucial nature of developing a comprehensive response.

Water, food and energy are the "first responders" to the changing climate, so people and communities feel the effects of change almost immediately. Rising temperatures are inexorably reducing icefields, glaciers and winter snowpacks, which in turn affect the timing and levels of river flows, meaning less water is reliably available in the summer months for food and hydropower production. Intense heat affects crop production – corn yields decline rapidly in temperatures exceeding 30°C, even for only one day. Multiyear droughts are projected to become common across the US Midwest and will have a major impact on agriculture, which accounts for over 70 per cent of all human freshwater consumption. Yet water use is not being reduced in most agricultural areas. Water is cheap; its use is abused due to generous subsidies and the powerful political lobbies that thrive on those subsidies. We now know that our current water-use practices are completely unsustainable. Not only is there not enough water to carry on as we have in the past, but the water we have always relied on may not be available in the future.

Even today's best hydro-climatic models reveal only a glimpse of how critical water is in the construct and maintenance of our Earthly reality at any given time. Life is made possible by all the ways in which water reacts with nearly every element in the physical world, and if a single defining factor in the Earth system changes, all of the other biogeochemical parameters also change. If global temperature changes, for example, an entire new geometry is created around that change. The most frightening discovery of this young century is that this is exactly what is happening worldwide, right before our very eyes.

Rising mean temperatures have begun to change a vast array of visible and invisible parameters that define the very foundation of the world as we have come to know it – at least how it is defined by water. These impacts on the global water cycle are causing what hydrologists call a "loss of stationarity," or "non-stationarity," which means a loss of relative stability and predictability in the water cycle. Stationarity is the notion that natural phenomena fluctuate within a fixed envelope of certainty that has allowed us to be pretty confident when it comes to predicting and managing the effects of weather and climate on our energy systems, cities and food systems. Engineers and others working in construction and development planning also rely on this predictability to define infrastructure safety standards.

The US National Academies of Science published a report in 2011 entitled *Global Change and Extreme Hydrology: Testing Conventional Wisdom*. It confirms how serious the loss of hydrologic stationarity could be if current trends persist. It concludes that, "continuing to use the assumption of stationarity in designing water management systems is no longer practical or defensible."

At a November 2014 meeting at the World Bank, the focus was on the real urgency in better communicating to others how our hydro-climatic circumstances are changing. The long-term hydrologic stability of the climate we experienced in the past, however, will not return during the lifetime of anyone alive today. What we haven't understood until now is the extent to which the fundamental stability of our political structures and global economy are in part predicated on relative hydrologic predictability. As a result of the loss of relative hydrologic stability, political stability and the stability of our global economy in some regions in the world are now at risk. This issue has become complex.

CLIMATE DISRUPTION AND DE-DEVELOPMENT

A recent map of interconnections between various economic, environmental, geopolitical, societal and technologic risks associated with the failure to effectively and meaningfully adapt to climate change was presented by the global insurance giant Munich Re at the World Economic Forum in Davos in 2015. The Munich Re map illustrates cascading effects of the failure to adapt to hydro-climatic change. On a global scale, failure leads first to greater vulnerability to extreme weather events, food crises, water crises, large-scale forced migration and further human-made environmental catastrophes, which in turn lead to accelerating biodiversity loss and Earth-system collapse. The failure to adapt also has devastating effects at the national level, where it can generate fiscal crises, unemployment, profound social instability, the failure of national governance, internal interstate conflict and terrorism and cyber-attacks, resulting in ongoing state crises leading to potential collapse. These risks are not theoretical. They are already happening.

According to a 2015 report by the UN, "Water in the World We Want," hydro-climatic destabilization is now a major threat to advancing development. Extreme weather events are now, in fact, seen to be reversing development in some regions. An example of this occurred in Pakistan in 2010 and 2011, where major floods happened, caused by heavy rainfall during the monsoon period. Land-use changes had altered natural drainage patterns and river flows, aggravating flood risk. More than 2,500 people died and 27 million people – three-quarters of the entire population of Canada – were displaced. The economic losses were estimated at US$7.4-billion. The country's development has been reversed, set back perhaps by decades.

Reports by both the UN and the World Bank, such as *Turn Down the Heat: Confronting the New Climate Normal*, make it clear that we now have to starting thinking the unthinkable. The unthinkable, of course, is that these kinds of events might reverse development here in North America, thereby reducing or threatening prosperity.

In fact, this is already happening. In 2014 the Weather Channel published a list of 50 US counties identified as being at high risk for weather, climate and natural disasters that damage and destroy homes and put lives and livelihoods at stake. Who would want to move to such places? How can such places sustain their tax base? How can such counties reverse de-development? Climate-related de-development is already happening in Canada also. We just don't see it that way yet.

We now confront a perfect storm in which insatiable demands for and on all critical elements of the nexus are not only competing with each other but their collective availability is being affected by the relentless drawdown of ecosystem services, the loss of biodiversity, and the implications of non-stationarity for hydrology and temperature regimes.

Policy-makers have coined the term "wicked problem" to define this predicament; we cannot solve this problem through traditional means, because there is no central authority accountable for dealing with it – it affects all nations and all levels of government. Furthermore, time is of the essence: the longer decisions are put off, the more severe the consequences. Unfortunately, humans tend to defer solutions to problems that are not seen to be immediate threats, because of all the short-term issues that need to be resolved. In democratic governments, politicians tend to focus on short-term policies affecting the election cycle, to the detriment of dealing with systemic and longer-term challenges.

But how are ordinary people taking and demanding action? Members of the political elite are absorbed in self-interested efforts to protect their investments in the very way of life that is threatening the nexus: fossil-fuel infrastructure, industrial-scale agriculture and constant demand for cheap water and increased supplies. But changes to the hydro-climatic regime are starting to render places that were formerly attractive for settlement and economic activity uninhabitable. Individuals are taking their own steps in shifting to renewables: installing solar panels, switching to drought-resistant gardens, buying electric cars. Entrepreneurs are responding with innovative technologies like batteries that store solar power overnight. A quiet revolution is under way, and it will become louder as the costs of inaction become increasingly severe.

NEW POLICY FRAMEWORKS

More attention is being given to dealing with the nexus as the world wakes up to the crisis. The year 2015 witnessed the development of several international policy framework agreements that collectively will have important bearing on the response to the nexus challenges sketched out above.

One such conference – resulting in the Sendai Framework for Action 2015–2030 – was held in Sendai, Japan, in March 2015. This agreement deals with reducing the impacts of natural disasters through all phases of the risk-reduction cycle: prevention, prediction, response and recovery. The conference built on the original Hyogo Framework for Action, created in 2005, which had gone some distance toward mitigating disasters around the world. However, the pace and magnitude of disasters has increased since then, and further work and investment will be required over the coming 15 years of the agreed action plan. The focus is on four priorities: a better understanding of risk; strengthened disaster-risk governance; increased investment; and "build back better," to ensure that investments in rehabilitation, recovery and reconstruction are more sustainable than before.

Another new policy framework was created when

the United Nations updated its set of Sustainable Development Goals in September 2015, upon the expiry of the original Millennium Development Goals of 2000. These goals will guide social, environmental and economic policy for decades to come. UN University's *Water in the World We Want* (February 2015), noted above, reports that unmet water goals threaten many world regions and form a barrier to key, universally shared ambitions, including stable political systems, greater wealth and better health for all. According to this report, the skilled and thoughtful management of water is the fundamental precondition of sustainability.

The report clarifies that the crisis is not that there is not enough water on Earth to meet all human needs; it is a crisis of there not being enough water where we want it, when we want it and of sufficient quality to meet human needs. But these needs have to change. Parts of the developed world, notably Australia and California, are already regulating to reduce water use. Adaptation in the nexus starts with policies and practices geared toward water conservation.

The February 2015 report also recognizes that we are currently experiencing an era of unprecedented change, especially in the manner and rate at which water moves through the global hydrologic cycle. Changes in the composition of the atmosphere – and thus the warming of climate – and the hydrological cycle mean greater impacts from deeper and more persistent droughts, more damaging floods and other extreme climatic events, which in turn affect the ways in which we live. Water-related disasters, climate disruption and social and economic development are inextricably linked.

Water security and climate security are inseparable: one is implicit in the other and both are essential for sustainable water development and management. According to James Lovelock, achieving water and climate security in the Anthropocene demands that we decide what hydro-climatic steady state we want and then set self-regulation toward achieving that state. One of the most immediate and cost-effective ways to begin do this is to invest in thoughtful, forward-thinking management of total water and related natural resources.

We lost international momentum in achieving the Millennium Development Goals with respect to water. If we don't regain that momentum and get moving on these matters now, population growth, economic instability, Earth-system impacts and climate disruption may make it impossible to ever achieve a meaningful level of sustainability. If this happens, we face stalled or reversed development: more people in poverty and greater national insecurity over water and water-related climate issues with the potential to create more international tension and conflict.

In responding to the urgency and the opportunity of finally getting sustainable development right, in September of 2015 the United Nations announced a new framework for global action. *Transforming Our World: The 2030 Agenda for Sustainable Development* promises to be the most comprehensive and inclusive effort to positively change the world in all of human history. This may well be the most important thing we have ever done for ourselves and for our planet. It is nothing less than a charter for people and the planet for the 21st century.

Theoretically, all the elements required to create sustainability are included in the agenda. The great challenge and urgency is to make these goals and targets a priority at the national level. This does not by any means suggest starting over; what it means is building on and focusing existing planning instruments and sustainable-development and resilience-enhancement strategies and technologies.

There are 17 goals set out in *Transforming Our*

World. In one way or another water plays a role in 14 of them. Goal 6 pertains specifically to water: to achieve universal and equitable access to safe and affordable water and sanitation for all. The goal also aims to improve water quality by reducing pollution, halving the proportion of untreated wastewater and substantially increasing recycling and safe reuse globally. Another aspect of this goal is to protect and restore water-related ecosystems in part by implementing integrated water-resource management at all levels, including transboundary basins. Another is to create livable, resilient cities. Can we do all this by 2030? Possibly, but it is not going to be easy and it is not going to be cheap.

If we want a sustainable world by 2030, we have to catch up with, and reverse, global poverty before population growth and Earth-system changes further destabilize our already weakened global economic outlook and make implementation of sustainable development goals unaffordable yet even more imperative. Development will not have to be simply sustainable. It will have to be restorative. The balance between environment, human security and economic viability will need to be articulated in a manner that holds all nations accountable for helping one another achieve the highest global standard for sustainable development, does not tolerate compromise, yet provides flexibility on the mechanisms by which to achieve those outcomes.

Another important global event in 2015 occurred in December, at the 21st UN Framework Convention on Climate Change (UNFCCC) Conference of the Parties (COP 21) in Paris. The goals of COP 21 are to negotiate a new international agreement on carbon-emissions reduction and a renewed commitment to adaptation to the changing climate. There will also be an increased commitment to meet the Green Climate Fund target of US$100-billion a year by 2020 to assist poorer nations to implement these goals. This new international agreement, which includes both developed and developing nations, will replace the defunct Kyoto Protocol, which was limited only to developed countries. We explore the full implications of this convention for Canada in the next chapter.

RESILIENCE, RESTORATION AND THE NEXUS

A significant component of the new policy framework defining the nexus between water, food, energy and climate is a recent focus on adaptation. Adaptation to the changing climate means accepting that hydro-climatic change will have profound effects on individuals, communities and natural systems.

Resilience is closely associated with adaptation. The Rockefeller Foundation defines resilience as "the capability of human and natural systems to survive, adapt and grow in face of shocks and even transform themselves when conditions permit." In the *Age of Consequences*, Courtney White defines resilience in ecological terms as "the capacity of a plant or animal population to recover from disruption and degradation caused by fire, flood, drought, insect infestation or other disturbance." Additionally, White notes, resilience describes a community's ability to adjust to incremental changes such as those brought about by changing precipitation patterns, warming temperatures and more frequent and intense extremes in weather conditions.

There are three levels of resilience: survival, adaptation and transformation. The first – survival – is the current "coping" model: institutions respond to catastrophes by resorting to the status quo. Floods occur, dikes are mended, houses are repaired and life carries on as usual until the next flood. The second level, adaptation, is a higher-order response that involves anticipatory planning and action: to carry forward the flood example, people might plan ahead to avoid or reduce future flooding by strengthening dikes or moving houses off the flood plain. The third level is transformation, which results in fundamental and permanent shifts in economic and governance systems. For example, areas becoming vulnerable to frequent floods may be permanently returned to nature, restoring ecosystem services in order to increase resilience to flooding. Economic incentives could be provided to encourage people to move and for governments to expand the range of policy response from engineering solutions to new approaches to land-use planning. Sustainable development must be redefined as restorative development.

In all cases, response to change should ensure a focus on emissions reduction and increased carbon absorption. Adaptation that increases emissions or damages ecosystems is really "maladaptation."

In order to achieve adaptation and transformation effectively, we must establish better links between the people working on the frontiers of hydro-climatic change: planners, engineers, architects and scientists. Practitioners will have to develop new codes of practice and standards to adapt to changing probabilities and magnitudes of droughts, floods, intense wind and rainstorms, and many are already beginning to address such changes in practices through renewed codes and standards of performance.

Enhancing existing social, environmental and political resilience, and creating long-term economic advantages for Canada in the face of hydro-climatic change, will require building a better bridge between science and public policy. Canada can capitalize on its already existing water wealth by identifying and acting on water's growing value to the national economy and to the future economic prosperity of the country at a time when water security is under threat in many other parts of the world. Scientists can transcend current jurisdictional fragmentation over water management across the country by translating research outcomes into sound and timely water-, food- and energy-management advice that should be offered simultaneously to engineering and related professions, industry, agricultural producers, municipalities, watershed councils, provinces and their federal agency counterparts.

Continuing to manage the nexus as we do today will bankrupt us. We have no choice but to visualize a different future and then realize that vision, knowing that we are not going to make better decisions by just being smarter at what we do today. If we don't confront the value system that created our problems in the first place, we will fail.

TWO

Warnings Repeated: The Intergovernmental Panel on Climate Change and the Global Nexus Narrative

The science explaining hydro-climatic change has become unnecessarily controversial, mainly because vested interests have funded some experts to select from the literature misleadingly biased analyses in order to cast doubt on the need for immediate action. The controversy is in part due to the inherent nature of science: scientists are always searching for verification of results and always finding more that is unknown than known, which limits professional experts from making definitive claims, especially pertaining to complex atmospheric and oceanographic systems.

Fortunately, the United Nations established an independent Intergovernmental Panel on Climate Change (IPCC) in 1988 to provide the highest globally accepted standard on scientific analysis of climate change. After the IPCC's first assessment report, the United Nations drafted its Framework Convention on Climate Change (UNFCCC). The mandate of the UNFCCC is to stabilize greenhouse gas emissions into the atmosphere to a level that would prevent dangerous anthropogenic change.

Since 1990 the IPCC has produced five assessment reports on the science of climate change. The most recent one was finalized in 2014 and is a synthesis of three interim reports. The first of those, on the physical science of climate change, was released in September 2013, followed by a second entitled *Impacts, Adaptation, and Vulnerability* (March 2014), and the third, *Mitigation of Climate Change* (April 2014). The synthesis report, released in November 2014, provides the scientific and policy basis for developing a global and legally binding treaty for reducing carbon emissions at the UN Convention on Climate in Paris in December 2015.

The IPCC undertakes a rigorous review of all the relevant science affecting the changing climate when preparing its reports for the UN. The panel does not carry out independent research; lead scientists carefully analyze peer-reviewed scientific literature as well as "grey literature" (climate model results, government documents and industry and non-governmental organization reports). Thousands of peer-reviewed science documents, together with thousands of comments from experts, are reviewed before the final reports are drafted.

In its 2014 summary report, the IPCC concluded unanimously that human influence has been the dominant cause in observed global warming since 1950, and that this warming of both the atmosphere and the oceans is unprecedented for over 1,400 years. Further, most of the warming is now occurring in the upper layers of the oceans, much less in the atmosphere. This absorption of heat by water accounts in large part for the apparent slowdown in the recent temperature increase in the atmosphere, which the climate skeptics have jumped on as proof that warming has ceased.

The concentration of carbon dioxide in the atmosphere has now reached an average of 400 parts per million, a level unprecedented on Earth for over 800,000 years. This gas is by far the most predominant of the greenhouse gases – the group of atmospheric chemicals that have a warming effect on the atmosphere as they are released. The other GHGs include methane (CH_4); nitrous oxides (NH_3); various hydrofluorocarbons (HFCs) and water vapour. Each of these gases has a distinct effect on atmospheric warming, so to simplify for analysis, scientists have developed a formula that expresses GHGs' warming effect as equivalent to CO_2, known as "CO_2 equivalent," or CO_2e.

CARBON BUDGET

In 2009 the political heads of UN member states formally accepted the IPCC target to limit the average global temperature increase to 2°C by stabilizing global GHG emissions. Currently, the increase is 0.85°C on average globally above pre-industrial times. The changes in climate that will occur through 2°C of warming will have major impacts for human society but should still allow life to continue, provided there are significant investments in adaptive measures. If warming increases over 2°C, and especially over 3°C, the IPCC predicts that the impacts on the crucial elements of the nexus will be severe and irreversible, threatening human systems and even existence on a global scale.

It has taken 150 years to release half the accumulated GHG emissions we can afford to emit if we are to keep the average temperature rise to within 2°C. It will take only about 25 years to release the other half at current rates of emissions. Many of the existing emissions are already "locked in" by economics and physics. To pay off initial capital investments, major carbon-producing infrastructure such as oil pipelines, coal-fired thermal plants and major transportation facilities require 40 years of productive life. It can take emissions up to 30 years from the time of release to reach the upper atmosphere and have full impact on climate. Consequently, decisions already made, or made now, on such infrastructure will likely lock in emissions and their impacts for almost one hundred years. Even if new sources of carbon were to be curtailed immediately, there would still be an increase of carbon release from existing infrastructure for several decades, resulting in a temperature rise of over 2°C.

The concentration of CO_2e in the atmosphere is now an average of 400 parts per million (ppm). The maximum allowable concentration of CO_2e set by the carbon budget established by the international community is 450 ppm. To achieve concentrations in the range of 450–500 ppm, total annual global emissions would have to be reduced by 1 per cent annually until 2030 and to zero by 2050.

PREDICTIONS

The Climate Action Tracker Consortium is a group of research institutions keeping tabs on each of the GHG emission target submissions by UN countries for the Paris conference in December 2015. The consortium has warned the UN that, as of mid-September 2015, the proposed emission targets (known in UN jargon as Intended Nationally Determined Contributions) of the 60 countries that have submitted thus far will fail to limit global warming to 2°C. This is despite some encouraging signs that world leaders were waking up to the need to take more aggressive action on carbon emission reductions. In a groundbreaking agreement in November 2014, the two largest carbon-emitting countries, the US and China – who together account for over 40 per cent of global emissions – committed to a significant shift in climate policy. The US administration set a target for reducing emissions by between 26 and 28 per cent below 2005 levels by 2025, significantly more ambitious than its previous target of a 17 per cent reduction. The Chinese government for the first time committed to capping its total emissions by 2030 and reducing them thereafter.

To illustrate the challenges of meeting this commitment, consider China, which dug up 3.87 billion tonnes of coal last year, more than enough to supply the next four largest coal users – the United States, India, the European Union and Russia – for a year. Not only does this action represent a huge contribution to global GHG emissions, but it also directly leads to the death of more than six million workers through lung disease and many more affected by severe air pollution in Chinese cities.

In a communiqué released by the G7 countries – Germany, France, the UK, Italy, the US, Japan and Canada – at a meeting in June 2015, leaders pledged to decarbonize their economies "during the course of the 21st century." For the first time, global leaders pledged to total elimination of carbon-based energy. This is part of a growing trend to make commitments for total phase-out of pollutants rather than percentage reductions, which are difficult to monitor and be held accountable for. The ball appears to be rolling.

The prospect of a legally binding agreement emerging from Paris to limit future carbon emissions has already had an impact on the fossil-fuel industry. At this writing, the G20 countries (representing the industrial and emerging world) have launched a probe into the potential financial risks posed by imposing such limits on costly investments in coal, oil and gas that may never materialize. This follows a similar probe undertaken by the Bank of England to review the implications on fossil-fuel investment if the UN target of limiting global temperature rise to 2°C becomes binding. The majority of proven reserves would be "unburnable," authors Christophe McGlade and Paul Ekins say in their 2015 *Nature* article, leading to what are termed "stranded assets" with an estimated value in the range of US$28-trillion.

The Bank of England's review has resulted in some large investors either pulling fossil-fuel equities from their portfolios or seriously considering it. If the UN-sponsored probe noted above were to indicate a real and increasing risk to future investments in fossil fuels, the associated disinvestment movement would grow from a trickle to a flood.

A second item on the agenda of the Preparatory UN Conference held in Lima, Peru, in December 2014 was to generate a Green Climate Fund of US$100-billion a year by 2020. This fund will enable poorer countries to adapt to the changing climate with money donated by developed countries. To date, only $10-billion a year has been committed to the fund, woefully short of what will be required by 2020 and beyond as the carbon budget is breached.

Even the UN has indicated that the fund will be inadequate to deal with the scope of changes to climate and hydrology over the coming decades, even if everyone paid in.

WHY SOILS MATTER

Like Earth's atmosphere, soil is an amalgamation of living things over time. Good soils are as porous and microscopically alive as a coral reef and just as crucially central to the vitality of the larger ecosystems they bring into existence and support. The bacteria, fungi, protozoa, nematodes and micro-arthropods, and all the creatures that rely upon them, such as earthworms, beetles and voles, form the soil food web. This web is the foundation of life on land on this planet.

But we are destroying our soils through deforestation, overgrazing, monoculture and industrial agribusiness. Worldwide we try to make up for diminished natural soil health by replacing it with artificial fertilizers, which wash away with depleted soils and contaminate water in floods that could themselves be prevented by restoring those soils. Compacting of soils has also become a global problem in that it does not permit soils to absorb and retain water over the longer term. A modelling study conducted by Yadu Pokhrel and Naota Hanasaki and colleagues published in *Nature Geoscience* in 2012 demonstrates that nearly half of current sea-level rise can be attributed to increased runoff over agricultural lands whose soil has been compacted to an extent that they no longer have the absorptive capacity they had when healthy.

In the best, uncompacted soils, microbial and fungal aggregates become tiny subsurface dams, each capturing a precious reservoir of water. Micro-organisms move around on the thin film of water that forms between and within the soil aggregates. Healthy soil, rich in micro-organisms and their aggregates, holds water like a sponge, releasing it slowly to plants as well as to aquifers, streams and rivers. In combination, these effects even out the release and availability of water throughout the year. Healthy soil is now seen as the best protection for crops during a drought, as well as the best protection from floods. In her book *The Soil Will Save Us: How Scientists, Farmers, and Foodies are Healing the Soil to Save the Planet*, Kristin Ohlson reports on the long-standing successes of American soil health experts in underscoring the value of healthy soils in controlling floods and droughts. In a chapter called "Letting Nature Do Its Job," Ohlson cites a 2012 test that demonstrated that deep, healthy soils can absorb up to 20 centimetres of rain an hour without flooding.

Soils rich in micro-organisms will also filter out pollutants, eventually draining pure water into an aquifer or stream course. Healthy soils remain humanity's first and foremost water purification system. Rebuilding soils as a means of enhancing natural processes of water purification is smart urban planning. In fact, in the last two decades, two hundred cities in 29 countries have forgone building expensive new water treatment plants and instead invested in watershed restoration that filters water naturally while at the same time enhancing flood protection.

Soils not only grow forests, supply food and absorb and purify water, they also store carbon. In *The Potential of U.S. Grazing Lands to Sequester Carbon and Mitigate the Greenhouse Effect*, and in many published papers and presentations to the U.S. Congress, legendary soil scientist Rattan Lal reports that we have already lost as much as 80 billion tons of carbon from our soils through inappropriate agricultural practices and short-sighted land use. All current IPCC warming projections are based solely on the effect of increased CO_2 in the atmosphere resulting from GHG emissions. They do not as yet take ecological responses to this warming into full account. It appears, however, that these feedbacks could be substantial.

Dr. John Harte is a professor in the Energy and Resources Group and Department of Environmental Science, Policy and Management at University of California at Berkeley. In an interview with the legendary American ecologist Paul Ehrlich in a book entitled *Hope on Earth: A Conversation*, Harte summarized the outcomes of research conducted over a period of nearly three decades aimed at measuring the ecosystem feedback effects of warming of 2°C on alpine soil plots. Harte told Ehrlich that one of the most important effects he observed in his heated plots was that they lost about 25 per cent of their soil carbon. From this, Harte reasoned that when such plots are heated, the carbon in the soil burns off as carbon dioxide. This, Harte surmised, had real implications in terms of global warming. If you examine all the world's soils to a depth of about a foot, he said, what you find is about four times more carbon sequestered in our planet's soils than there is now in the atmosphere. Harte calculated that if you release a quarter of that into the atmosphere, that amount would be equal to the amount of carbon already in the atmosphere as a result of burning fossil fuels. In other words, if we warm the world's soils by 2°C, the amount of carbon dioxide in the atmosphere could double from its present 400 parts per million to 800 ppm.

The latest IPCC projections indicate that we are now on pace for a 4°C global mean surface temperature increase, which translates into a 5–6°C increase on land. This would result in heat waves in China 6–8°C hotter than ever experienced before; heat waves in Europe 8°C warmer than in 2003, when 70,000 people died of heat effects; and the warmest days in New York City becoming 12°C hotter than at present – all of which suggest an urgent need to take soil restoration seriously now.

RESTORATIVE AGRICULTURE

Keeping and getting carbon into the soil may be one of humanity's most important climate change mitigation strategies. It is no longer enough that agriculture just be sustainable; it has to be restorative. There are huge rewards for thinking of agricultural practices in terms of their regenerative value.

Kristin Ohlson's *The Soil Will Save Us* documents a variety of scientists' findings related to soil and mitigating climate change. New Mexico

State University molecular biologist David Johnson and his colleagues John Ellington and Wes Eaton recently demonstrated, for example, that even in arid regions soil carbon can be increased by 67 per cent and water-holding capacity by 30 per cent while at the same time increasing productivity through effective soil management and by using cover crops. In North Dakota, farmer Gabe Brown, an expert on restorative agriculture, has demonstrated that even a cover crop consisting of two to three species results in a 90 per cent reduction in soil sediment runoff and 50 per cent less fertilizer runoff into the watershed while at the same time sequestering more than one metric ton per hectare. This practice alone, if applied widely in agriculture in Canada, could conceivably absorb at least 5 per cent of current GHG emissions.

Rattan Lal believes that through restoration, 3 billion tonnes of carbon could be sequestered out of the atmosphere into Earth's soils, reducing atmospheric carbon dioxide concentrations by three parts per million each year. In other words, through astute soil conservation, we can conceivably slow or even possibly reverse warming. But restorative agriculture must be seen as only one element of a larger sustainable restorative development conversation our society has to have with itself to manage the crisis in the nexus.

SUSTAINABLE DEVELOPMENT = RESTORATIVE DEVELOPMENT

We have begun to realize that we need to reframe what we know about what is happening to the Earth system in new ways that may positively affect behavioural change leading to meaningful action on climate disruption and that the impetus for change could be the idea that, in the context of climate change, sustainable development means restorative development.

To restore Earth's stability we have to harness the power of informed choice by giving people the capacity to address the climate threat on a personal level. Human beings are wired mentally for a stable climate. We have to accept, however, that our climate is no longer stable and that this poses a huge danger to our future. We can no longer justify our own inaction by focusing on why others don't act or refuse to act. Instead, we should focus communications attention on the vast number of people in "the floating middle" who are politically influential and open to being informed by scientific evidence.

If we want to see real change, we need to translate

what we know into action: identifying and supporting those places where action is most likely to occur. A place to start is in progressive cities that have already begun to advance quickly in the direction of both mitigation and adaptation. Our sense of social responsibility needs to be informed by values that transcend self-satisfaction, conformity and the status quo; in-group association; and partisanship and ideology. We need to appeal to deeper instinctual dispositions such as caring for our children and coming generations, as well as that oldest and deepest human trait which recognizes that in order to survive and prosper as individuals we have to account for each other. We need to demonstrate that we are all in this together. If we want to restore the foundation of our prosperity, we need everyone in. Restorative development and regenerative agriculture are within our means. What is needed, however, is a sense of urgency. We would be wise to derive that sense of urgency from what we see happening now in places like California.

WATER SECURITY AND HUMAN VULNERABILITY: LESSONS FROM CALIFORNIA

In his book *American Exodus: Climate Change and the Coming Flight for Survival*, Giles Slade argues that this conflict is already happening and that people are on the move. Slade estimates that about 3.5 million middle-class citizens have left drought-stricken California since 1993, in part because of water restrictions and increasing property damage associated with wildfires. They are being replaced by large numbers of Latin American citizens, themselves fleeing from a range of challenges in Central and South America. These population shifts are already having enormous political, economic and social implications, yet we are witnessing only the tip of the iceberg.

People who become desperate as a result of climate change will take enormous risks to find refuge. Slade argues that the relatively improved availability of fresh water in Canada, plus the stable political institutions, will become an enormous attraction for many Americans as the Southwest faces increasing drought over the coming decades. At this writing, climate disruption has already caused economic hardship and social and political upheaval in California. What lessons can we learn from California's situation that might help us understand the trouble brewing at the nexus of water, food, energy and climate in Canada?

California and Canada have a lot more in common than we might have thought. California has a long history of decades-long droughts and megafloods. This is also the case on the Canadian prairies. As the great salt lakes of the American Southwest clearly demonstrate, abundant water can disappear very quickly in a region that is highly sensitive to climate change. In the 12th century the flows of the Colorado River are estimated to have been 85 per cent lower than the 20th century mean upon which the Colorado River apportionment compact between the western states was founded.

Droughts of such extended duration have occurred in the American Southwest that they are thought to have ended the highly adapted Pueblo civilization. At a certain size, populations become too big to move, and immobile populations can find themselves in a world of hurt if climate disruptions persist. Skeletal remains from the end of the Pueblo period offer clear evidence of suffering, malnutrition, starvation, disease, high infant mortality and dramatically shortened lifespans.

And of course, there is a direct link between drought and fire. When surface streamflow declines, groundwater depletion accelerates and trees suffer. When weakened by drought and insect pests, forests catch fire. There is also, as we have seen in California, a direct link between drought and decreased mountain snowpack.

For each 1°C rise in temperature, the snowline in the American West rises more than 150 metres. With an increase of 3°C, the snowline rises 500 metres. The Sierra Nevada will lose half of its snowpack with 3°C of warming. The Weather Network reported in September 2015 that the snowpack in the Sierra Nevada shrank to its lowest in 500 years. There is also a direct link between more precipitation falling as rain as opposed to snow in winter and the potential for flooding due to the increased likelihood of warm rain on snow events in both spring and fall.

The California drought further calls into question the validity of the statistics we use to design infrastructure and protect people from exposure to harm caused by meteorological disasters. What is generally assumed to be a one in a hundred year flood, for example, is based on what we have observed over the past 50 years and is therefore in the context of the larger hydrometeorological record.

We need to judge such events by the paleo record, which establishes a whole new baseline for what the flood of record actually means.

Floods in California today are nothing compared to what have occurred in the past; and the state is long overdue for another big one. California's biggest floods have been caused by storms popularly known as the Pineapple Express, associated with El Niño events caused by a reversal of currents in the Pacific Ocean. We now know that these storms are actually atmospheric rivers: huge currents of water vapour aloft, drawn across the Pacific on near-hurricane-force winds. When moderate in scale, these bring badly needed water to a dry state. When larger, however, they cause flooding of a magnitude we can hardly imagine.

Though it may seem counterintuitive, California is now more vulnerable to flooding than ever. As we have observed, groundwater overdraft has caused land subsidence of up to 9 metres in parts of the state, making these areas even more vulnerable to flooding. California is a flood disaster waiting to happen. With each drought and flood the costs mount to the economy and to collapsing ecosystems. The U.S. Geological Survey has published a new emergency preparedness scenario called ARkStorm

(Atmospheric River 1000 Storm). The scenario demonstrates that if struck by a one in a thousand year atmospheric river event like the one that happened in 1861–1862, California would lose one quarter of all homes, with projected damage of US$725-billion. To transform in the face of this change, Californians have to stop building in flood plains and begin restoring natural wetlands.

It is important to recognize that what we have experienced so far with a global average temperature increase of 0.8°C is the relatively benign range of vulnerability we can expect in what people now refer to as the new normal. We have to transcend our short flood memory and create a new water ethic for our society. The days when we could simply expect to increase our demands for water are over. We must integrate supply management with water quality and flood protection and improve monitoring and prediction.

We all need to pay more attention to what is happening in places like California. California and the American Southwest appear to be facing a hydroclimatic double whammy: a cyclic return of drier conditions, combined with the effects of warming brought about by changes in the composition of the atmosphere resulting from our growing numbers and accelerated activities in the past century. From the California example we learn that the effects of rapid climate warming are never uniform. Deep and persistent droughts are followed by severe storms and flooding. This oscillation between extreme wet and dry conditions knocks out ecosystems, whether it occurs in California or on the Canadian prairies. Droughts of up to four hundred years in duration have overlapped from the Southwest into California in the past. Dr. David Sauchyn and colleagues Jodi Axelson and Jonathan Barichivich at the University of Regina have shown that droughts of similar magnitude have also occurred on the Canadian Great Plains.

The difference between the drought of the 1930s and the one from 1987–1992 in California, and the one that started there in 2014, is larger human populations and greater vulnerability. We can expect larger floods and protracted droughts now and in the future. We can – and likely will – adapt to hydro-climatic change, but the price will be high. As different regions adapt, however, they should aim to be places where people will want to live in a warming world.

ADAPTATION AND IMPLICATIONS FOR THE NEXUS

As climate continues to change with the locked-in GHG emissions already in the atmosphere, we can assume that more and more places will become inhospitable. This will lead to shifts in people's comfort zones, especially among those who operate farms and energy systems in the face of erratic water security. Risks of fires, floods and droughts will come to dominate social consciousness over the coming decades.

As already noted, global carbon emissions will continue to increase until at least 2030, and by then their effects on the oceans and atmosphere will be irreversible. In addition, there will continue to be changes in land use due to reclamation of increased areas for agricultural production and ongoing loss of forests and wetlands due to drainage. All these changes will reduce the natural capacity of ecosystems to store carbon. There is thus overwhelming evidence that climate change will have significant consequences across the entire range of the water, food, energy and climate nexus.

The more interconnected the nexus becomes, the more vulnerable it is to tipping points – events in which the entire system shifts to a new state. Once tipping points occur, it is usually very difficult and often impossible to shift back to the earlier state. The water, energy, food and biodiversity systems that all life relies upon will therefore become less secure and more prone to widespread, cascading change. The most crucial factor is water, as its reduced quality and availability has serious implications for energy and food production. Likewise, agricultural and food processes are highly reliant on affordable and stable energy sources. The more that ecosystems are modified, the less resilient they become, the less able to adapt to external stresses without losing function or structure.

Adaptation to variable weather in agriculture is not a new concept. Farmers have always incorporated weather-related risk management in their practice, and policy-makers have supported insurance programs to assist in coping with variability in weather for decades. However, what we now face are structural, regional and permanent changes to the climate. In short, farmers are adapted to changes in the weather but not to changes in the climate. We thus require a coordinated approach, involving both the public and private sectors, to adapt to this pending crisis in the nexus.

In its previous reports, the Adaptation to Climate

Change Team (ACT) at Simon Fraser University has identified the following key factors that illustrate the need for a comprehensive approach to adaptation across the nexus. Projected lengthening of the growing season in Canada will allow northern expansion of warm-weather crops, so long as soil and water conditions are suitable and supporting infrastructure can be economically developed. Climate change is affecting regions from which Canada imports its foods, which means that Canada must increase production locally to compensate and enhance food security in the future. A much more coordinated effort should be undertaken to reduce waste across the entire food-supply chain. Today in Canada, between 30 and 50 per cent of food is wasted along with the water and energy required to produce and transport it. In a world facing the predicted severities of a changing climate, such waste cannot be tolerated. But this requires a fundamental shift in culture from a throwaway society to a conserving society where there is no such thing as waste.

Current food and energy production creates a market in "virtual water" – water embedded in food products that are transported overseas. In a prolonged drought, California will have to make choices between food export and food for domestic staples. For example, 70 per cent of the state's almonds are exported, yet producing them consumes twice as much water as basic staples such as cotton and tomatoes.

Even in drought-stricken California, old customs and attitudes die hard. On April 1, 2015, *The New York Times* quoted Representative Kevin McCarthy: "[W]e know that we cannot conserve or ration our way out of this drought." Agribusinesses are lobbying their state representatives to secure funding to increase supplies, even though no additional projects could be up and running within 20 years due to complex commission rules. "Water flows uphill, toward the money," as the old saying goes. Agricultural water supply has been massively subsidized for over 60 years in California. In some water-supply projects, farmers have repaid only 15 per cent of capital and operating costs in more than six decades. To make things worse, some farmers are not only given generous subsidies for water but also agricultural subsidies for growing commodity crops such as rice and cotton.

In addition, across the North American West, water rights are allocated on the basis of first in right, first in time. In other words, places with senior rights can abuse their water supplies while junior rights holders can be denied access even if they have developed much more efficient water systems. With handsome subsidies for water across California, there is little or no incentive to conserve or protect watersheds. Most of the water is used to irrigate relatively low-value crops such as alfalfa and pasture to support the meat and dairy industry – not the higher-valued vegetable and fruit crops. The human-caused disaster in California is due not only to the changing climate but also a long-embedded culture that undervalues and wastes water. The shift to a society that conserves water is just one that will have to happen in order to reduce the tension in the nexus over the coming decade.

We also need policy changes related to groundwater. In many agricultural areas of the western states, groundwater is being mined at unsustainable rates. This has so alarmed legislators that changes are already afoot. California passed the Sustainable Groundwater Management Act in 2014 in response to the multiyear drought which caused not only alarming groundwater depletion but also subsidence and collapse of some groundwater aquifers and contamination by seawater intrusion. The legislation empowers local agencies to develop groundwater-management plans, enables the state government to intervene if local agencies fail to do so, and allows the state to delay implementation of local plans if they involve complex interactions between groundwater and surface water rights. The groundwater plans allocate water among users for the first time and can limit or suspend such allocations if the drought emergency continues. All high-risk areas have to develop plans by 2020, lower-risk ones by 2022. The state will publish its estimate of total groundwater resources available for extraction on a sustainable basis by 2017. Though still in its infancy, this legislation is a benchmark in North America for groundwater regulation.

British Columbia too will regulate groundwater extraction for the first time, under the Water Sustainability Act, also passed in 2014 though not yet in force. The province is currently drafting regulations to establish which users will be regulated (only large-scale ones initially) and to set out monitoring requirements and specific instructions on beneficial use of the resource to reduce overconsumption. The regulations are expected to come into force in 2016. More sophisticated regulation of water for both consumption and ecology will be essential if the nexus is to become more resilient to the changing hydro-climatic regime.

Extreme weather events can now affect the global price of food staples like grain and rice. Food prices shot up in 2011 as a result of the drought in Russia, and in 2012 as a result of a similar drought in the southern United States. Heat waves of over 30°C, now increasingly common in areas where staple crops are grown, can markedly affect yields even if the wave lasts only a few days. The IPCC has predicted that yields of corn, soy and cotton will face declines of between 30 and 80 per cent by the end of the century. Land degradation will contribute to these declines, requiring more fertilizer and drainage controls, both of which will increase energy demands and negatively affect water supplies. Food quality is also impacted by climate change, notably for critical commodities – think wine, coffee and cattle forage; heat-induced stress can also affect milk production.

Fisheries impacts will also become more severe as the decades progress. The IPCC forecasts a 30–70 per cent decrease in fisheries yields by 2055 in some high-latitude regions, and a 40–60 per cent drop in the tropics, based on a 2°C warming scenario. With the likelihood of greater ocean warming, increasing acidity and the lack of international coordination of ocean fisheries, the prospects for sustaining fisheries at levels that come close to meeting the food demands of a growing population are extremely dim. Temperature increases in rivers such as the Fraser and the Columbia will further affect the survival of species like Pacific salmon. In the summer of 2015, 500,000 Pacific salmon returning to the Okanagan system in the Columbia basin were lost due to high water temperatures. Only more northerly watersheds are expected to be productive for fisheries after mid-century. People in southern British Columbia may have to focus more on aquaculture, but this is an expensive and energy-intensive prospect with significant environmental impacts. And we should be keeping close tabs on energy use.

Changing hydrologic cycles will have significant effects on existing hydroelectric projects. At higher altitudes, such as in the Canadian Rockies, glaciers, which provide critical flows in summer months, are expected to largely disappear before the end of the century. Winter snowfall will gradually be replaced by rains flowing through the reservoirs in spring and leaving much less water for the critical food-producing months in the summer. Conflicts will arise between those who wish to retain water flows in rivers and streams for food fish and those who wish to exploit flows to irrigate crops.

All these hydrologic changes will also affect electricity generation by existing hydro dams. Energy production might well be compromised by reduced flows in mid- to late summer, the very time when power demands for air conditioning and for pumping both surface and groundwater are at their highest.

Many coal-fired generating plants are located in regions where surface water will become scarce and groundwater resources will be depleted, such as the southern United States and parts of China. Coal-fired industry not only has a large carbon emissions footprint but also an insatiable thirst for water. We all need to switch from coal to more sustainable sources of energy in the near future – a major challenge, given its current predominance as an energy source in developing countries.

Over the past 50 years, humans have changed ecosystems more rapidly than in any comparable period in human history, largely to meet the growing demands of the elements of the nexus. This has resulted in a massive loss of biodiversity and destruction of the resilience of remaining ecosystems, which has compromised their ability to adapt to anticipated changes. Decision-makers simply do not account for loss of ecosystem services, which leads to inappropriate taxes and subsidies that encourage overexploitation and unsustainable practices.

In addition, disparities in income, corruption in many developing governance institutions, lack of transparency in decision-making and inadequate information constrain future management of ecosystems the world over. Achieving properly functioning governance institutions and supporting social systems is essential if we are to regulate use of ecosystem services.

The crisis at the nexus is a result of a combination of factors: a reduction in water supply, the changing energy balance and disruption in the food supply chain. Continual changes in climate, and frequency and severity of extreme weather events together with a loss of ecosystem services, have increased risks to all of these elements. To adapt to these changing hydro-climatic circumstances, we have to explore how governments, businesses and the public are dealing with this crisis in Canada.

THREE

Bringing It Home: The Nexus Narrative in Canada

As we stated before, there is no likelihood we will meet the target global carbon emissions reduction goals approved by the UN to limit global temperature increase to 2°C. There is too much infrastructure in place and too many carbon-based energy resources already committed for development for us to turn the tide in time. Even the IPCC has predicted that global temperatures will rise to between 3°C and 5°C by the end of the century. Canada and British Columbia are not, however, at a standing start with respect to policy changes to reduce impacts on the nexus. There have been important incremental changes in policy as a result of the growing awareness of the changing climate, supported by recommendations from ACT. British Columbia has been a leader in this regard.

Until 2015 the federal government's formal pledge to the UN was to reduce GHG emissions by 17 per cent below 2005 levels by 2020. To date it has relied on a regulatory approach to emissions reduction. It has not enacted promised regulations for the oil and gas sector, however. In 2012 Canada's emissions were 18 per cent above the target for 2020. But in May 2015 the Harper government announced a new target of 30 per cent reduction from the 2005 base by 2030. The Climate Action Tracker Consortium of research institutions tabulating submissions from UN member countries for the Paris conference states bluntly that the Canadian target is "inadequate." It is based in part on proposed new regulations to control methane emissions from the oil and gas sector, such as industrial leaks and gas flaring; reducing GHG emissions from natural-gas fired electricity generators; and new standards to limit emissions from the chemical sector (specifically from nitrogen fertilizers). The remainder of the overall reduction would result from more aggressive targets set by the provinces.

British Columbia has legislated a target reduction in GHG emissions of 33 per cent below 2007 levels by 2020. This target is built around a carbon tax introduced in 2008. British Columbia was the first jurisdiction in North America to introduce a tax on all carbon burned as a fossil fuel. The tax was increased gradually over five years and currently stands at $30 per tonne. This levy is offset by reductions in personal and corporate income tax so that it is largely revenue neutral. Although independent experts lauded the policy for reducing consumption of gasoline and home-heating fuels, the current government has decided to maintain the tax at its current level for the foreseeable future. The carbon tax has elicited international interest from policymakers who see the British Columbia model as a potential trendsetter.

The province also introduced a Clean Energy Act, which lists, among other objectives, the goal of having 93 per cent of the province's electricity come from renewable resources. However, the province recently altered the definition of "clean fuels" to include natural gas, provided it meets a high level of efficiency and waste heat is reused. (This redefinition coincided with the government's decision to invest massive efforts in development of a provincial liquefied natural gas [LNG] industry.) It also requires at least 5 per cent renewable content in gasoline and diesel and a 10 per cent carbon-reduction intensity in transportation fuels by 2020.

In May 2015 the federal government announced that it will develop a strategy to reduce methane emissions from natural gas fracking in northern British Columbia and along pipelines routes. And similarly the province has introduced legislation that will control, project by project, some of the emissions from proposed LNG plants for the North Coast.

The Pembina Institute, a group of energy experts, has stated that BC will miss its 2020 legislative target due to its promotion of LNG plants. To address this criticism, the province established the multisector Climate Leadership Team to provide recommendations on how to meet targets, improve government-to-government relations with First Nations regarding energy projects, establish adaptation approaches and collaborate with local governments on climate action plans. This team was to issue its first report in time for the Paris Climate Conference, based on public input. The provincial government was to finalize its updated climate action plan, based on the work of this team, in the spring of 2016.

Alberta has implemented an alternative financial model to reduce emissions. All major industry has to meet prescribed reductions in carbon intensity – the amount of carbon emitted per unit of production. As overall production increases, so do total carbon emissions. Where the regulated targets are not met, industry pays a levy – currently $15 per tonne – into a fund to support development of green technologies and energy efficiency. In June 2015 the newly elected provincial government announced that the levy will be increased to $20 per tonne on January 1, 2016, and to $30 per tonne in 2017. In addition, the levy will apply to industries that cannot reduce their carbon intensity by 15 per cent in 2016 and 20 per cent by 2017. The intensity target is currently 12 per cent. As the financial levy applies to only about 4 per cent of total carbon emissions in the province, it has limited utility. In July 2014 the Auditor General had criticized Alberta for not implementing its emission-control strategies, noting that emissions in 2012 were 50 per cent higher than in 1990. The provincial government promised to announce a more comprehensive emissions reduction plan before the Paris conference.

In April 2015 Quebec and Ontario announced the development of cap and trade policies to reduce carbon emissions. The Quebec target at the time of the announcement was 20 per cent below 1990 levels by 2020. In September 2015 Quebec strengthened its target to 37.5 per cent below 1990 levels over the course of the next 15 years. It estimates the cap and trade system will raise $2.8-million by 2020, which will go into a fund to implement its climate action plan. Ontario had originally set a target of a 15 per cent reduction below 1990 but recently announced that it would increase this to a 35 per cent reduction by 2030. Ontario phased out its coal-fired generating plants in 2014 and is committed to reducing its carbon content in transportation fuels by 10 per cent by 2020.

The cap and trade system is simple in concept, but in practice its application has been a failure. The provinces set a cap on the total carbon emissions by granting permits to the major emitters. The more efficient operators are in a position to trade some of their reductions below their permitted amounts to the less efficient operators who buy (or trade) some of this capacity, such that the overall cap is maintained. Over time the cap is steadily reduced to meet national carbon-reduction targets. In theory, as the cap is lowered the price for trading in carbon should rise, thus reinforcing the financial incentive. Again in theory, the cap and trade policy advocates believe that the overall plan is more efficient than a straight carbon tax.

But theory and practice do not match. Politicians implementing cap and trade policies the world over issue too many permits and even exclude some industries from the scheme altogether, industries seen to be "vital for the national interest" (for example, coal mines and power plants in parts of Europe). Europe prides itself on being a leader in carbon reduction, but it initially gave away too many permits and then got caught up in recession. The initial price of carbon was US$25 per tonne when the EU Emission Trading Scheme was introduced in 2005 but in 2015 was hovering under $5 per tonne. Politics makes polluting cheap.

In order to achieve real reductions in carbon emissions, the price of carbon has to rise significantly. In June 2015 the International Energy Agency pointed out that prices would need to rise to US$65 per tonne before power plants would switch from coal to natural gas. Similarly, carbon capture and sequestration technology, considered an essential tool in the global carbon-reduction strategy, will be viable only if the price of carbon rises substantially.

Many analysts argue that a straight carbon tax is more effective in driving carbon reduction than cap and trade even though it is much less flexible. But few politicians (British Columbia's excepted) will risk their seats by introducing a carbon tax when almost all political parties in the developed world shy away from new tax measures.

Almost all British Columbia municipalities have signed a Climate Action Charter that calls for plans to reduce carbon emissions in order for municipalities to be exempted from the carbon tax. Plans include bylaws to encourage energy-efficient buildings, wastewater recycling and bio-energy production as well as improvements to public transportation and developing bicycle infrastructure.

Most expert observers feel that Canada will not meet its UN target unless it brings in new policies, such as regulations for the oil and gas industry, some form of financial levy on carbon emissions, and further development of carbon capture and sequestration technology. The Harper government had remained committed to supporting large-scale infrastructure for exporting oil and natural gas, including pipelines from Alberta to both coasts and large-scale investment in LNG in British Columbia.

Here is the problem: effective mitigation cannot be achieved if each nation pursues its own interests independently. A tonne of carbon released from

Canada has the same impact on the climate as a tonne of carbon emitted from China. International co-operation is essential to mitigate GHG emissions effectively, but to date, consensus has eluded global decision-makers despite more than two decades of UN-sponsored climate meetings.

GLOBAL AGREEMENT ON CARBON REDUCTION

The UN conference on climate in Paris in December 2015 was yet another attempt to craft a globally binding agreement for reducing carbon emissions. The first such pact – the Kyoto Protocol – set a legally binding target for developed countries: a 5 per cent reduction in GHG emissions below 1990 levels by the period 2008–2012. That these target reductions applied only to developed countries was the protocol's death knell; by 2000, developing countries like China were emitting as much carbon as developed countries, and some objected to giving developing countries an unfair competitive advantage. Consequently the protocol was never ratified by the US, and Canada reneged on its ratification.

It is difficult to negotiate a global treaty. Issues of equity, fairness and justice arise with respect to both mitigation and adaptation. Developed countries have emitted most of the GHGs over the past 150 years and have already consumed about half of the atmospheric capacity to absorb carbon within the target set by the UN. As a result, developing countries that now emit almost half the global GHG emissions argue that developed counties should play a lead role in emissions reduction. Developing countries also assert that developed countries should pay the lion's share of the Green Climate Fund of US$100-billion a year by 2020.

The logjam that would require all countries to contribute to GHG emissions reductions was finally broken in Lima, Peru, in December 2014. Each country was required to submit its national targets between March and June 2015 so that the UN Secretariat could analyze them ahead of the Paris meeting. At this writing, about 66 per cent of global emission targets had been submitted. The most important is the new US commitment to reduce emissions by between 26 and 28 per cent by 2025 compared with 2007 levels. This is a significant increase from its earlier 17 per cent commitment and is based on the US administration's pledge to bring in regulations to reduce emissions from its massive coal sector by 32 per cent from 2005 levels by 2030. In September 2015 China further committed to introduce a cap and trade system to reduce its emissions from power generation, iron and steel and the chemical sector, starting in 2017.

In June 2015 the Roman Catholic pontiff delivered an encyclical letter that was the first such papal statement to be entirely devoted to climate and protecting nature's biodiversity. The following September Pope Francis urged all members of the UN to commit to renewed efforts to come to an historic agreement to curb carbon emissions and assist developing and underdeveloped nations to adapt to the changing climate.

All of these moves lent momentum to the preparations for the Paris conference.

ADAPTATION, VULNERABILITY AND ADAPTIVE CAPACITY

Because national and global policies to reduce GHG emissions will not achieve the IPCC carbon budget target to keep global temperature rise to under 2°C, much attention has been given to adaptation strategies at all levels of government. As mentioned earlier, adaptation involves purposeful anticipation of change and careful planning to reduce the effects of change. Adaptive capacity requires the marshalling of significant resources – economic and technological – as well as skills in planning, engineering and creating social and behavioural change, and ensuring equity and fairness in decision-making.

In Canada, federal and provincial governments have begun to invest in adaptation measures. In 2011 the federal government announced a five-year Adaptation Platform for coordinating institutional, financial and knowledge resources for mitigating the effects of climate change across Canada. The federal government's pledge of $148.8-million can be leveraged by other levels of government and the non-profit sector to increase total investment. Though a significant investment is directed to predicting climate change scenarios for regions across Canada, much of this program is not specifically directed toward the nexus, though some of the projects that are relevant are noted below.

Biodiversity

ACT recognized from its outset that adaptation is closely linked to resiliency – the ability of systems (natural, social or technological) to withstand change and retain their functionality. The very first ACT report, *Climate Change Adaptation and Biodiversity* (2008), focused on the ability of natural systems to adapt to changing hydro-climatic conditions while sustaining their ecosystem values. This analysis was applied to British Columbia, partly because that province has the most diverse ecosystems in the country and partly because rapid development of natural resources there has reduced ecosystem resiliency.

In British Columbia, unlike in other provinces, the Crown owns approximately 95 per cent of the land and resources. Consequently, almost all resource sector development requires government approval. To provide a strategic road map for guiding such developments, the provincial government prepared 26 Land and Resource Management Plans across the province between 1993 and 2008. These plans established areas recognized as having such high environmental and biological values that they should be fully protected from resource development. Over the 15 years of planning, the size of protected areas increased from 6 per cent to 14 per cent of the province. For the rest of the Crown lands, the plans set out general principles for resource development that were in turn supported by legal standards laid out by numerous pieces of resource legislation, each enforced by a separate agency.

ACT noted that though individual agencies may meet their statutory requirements, the cumulative effect of many separate development actions in a single ecosystem like a watershed could lead to loss of ecosystem function. Accordingly ACT recommended that a single agency be responsible for all land and water decisions on Crown land. Such an agency was established in 2010 and is now developing an integrated decision-making model so that a single decider makes all rulings related to land and water in watersheds.

ACT further recommended that the government tackle the cumulative effects of multiple developments on single ecosystems, which can exceed nature's limits and bring about unintentional loss of ecosystem function. ACT proposed that initial attention be paid to ecosystems of higher value, those that are not functioning well and those located where there is highly valued resource-development potential. Pilot cumulative effect assessments are currently under way in high-risk areas to develop information requirements and improve decision-making tools. These pilots will be completed by 2021 and will provide a more robust approach to sustaining ecosystem resiliency in light of the changing climate as well as enabling well-designed resource development.

In May 2015 British Columbia's Auditor General

reviewed these pilot projects on cumulative effects as applied in the northwestern region of the province. She was critical of the length of time it took to implement the approach, given the multitude of potential developments in the region. She also criticized the lack of clear legislative authority to change the rules associated with the existing sectoral approach and emphasized the needs for clearer direction from government and for monitoring to support the new approach. The provincial government has responded by bringing implementation of the cumulative effects framework forward to 2016 for operational decisions and for strategic decisions such as plans and resource agreements and treaty negotiations with First Nations. Full implementation will be evaluated by 2021.

The federal government initiated a National Conservation Plan in 2014 to protect and restore key ecosystems across the country. The plan provides investments of $250-million over five years, with requirements for matching funds from provinces and other sources, focusing on three priorities: conserving biodiversity in land and marine ecosystems; restoring degraded ecosystems to support species at risk, clean water and wildlife; and working with non-profit organizations to increase conservation areas in and around communities and build public appreciation for nature. The plan supports Canada's commitment under the UN Convention on Biodiversity to meet biodiversity goals and targets by 2020.

Water and Watershed Governance

Water is generally undervalued in Canada, and as a result it is overused. ACT's 2012 report *Climate Change Adaptation and Water Governance* recommended strengthening national and regional water-conservation guidelines, allocating water to meet nature's needs, and in the long term establishing a new water ethic centred on reuse – making water fit for purpose and conservation. Some provinces and local governments have introduced pricing regimes that encourage more efficient use of water. Given the increased likelihood of droughts, more attention will have to be given to drought prediction and management.

British Columbia has implemented ACT's recommendations with its 2014 Water Sustainability Act, which enables the province to establish flows for ecological purposes before awarding new licences for water withdrawals. The Act also enables development of drought management plans which, when required, can reduce the amount of water withdrawn under licence to meet essential needs, thus retaining base flows in watercourses even during extreme droughts. In addition, the Act enables the setting of provincial water objectives for protecting water quality and flows and ecosystem needs. These objectives will be integrated with other land-based resource and environmental objectives under the cumulative effects framework noted above to provide a more comprehensive approach for making water and land-use decisions. All of these provisions will be introduced through regulations starting in 2016.

The Water Sustainability Act further provides for protecting future water requirements for farming through agricultural water reserves. The Minister of Environment can authorize these reserves through the approval of water sustainability plans based on extensive public consultation. Regulations supporting the development of such plans are scheduled to be issued in 2016.

Agriculture and Water

In 2013 the BC Ministry of Agriculture initiated a significant climate change adaptation program funded through *Growing Forward 2*, a five-year federal–provincial–territorial plan delivered by the

BC Agriculture and Food Climate Action Initiative. The program covers adaptation at both the farm level and the wider regional level.

The farm-level program supports innovative practices that have potential to help producers adapt to climate change. By September 2015, 12 pilot and demonstration projects were under way, with a focus on strengthening farm resilience, transferring knowledge across the sector and monitoring performance.

Regional agricultural adaptation strategies have been completed for the Cowichan Valley, the Corporation of Delta, the Peace and Cariboo regions and the Fraser Valley. A strategy for the Okanagan region is under development. Each of the regional strategies presents local climate change projections for the 2020s and the 2050s, anticipated agricultural impacts and recommended actions to deal with priority impacts. The most significant issues under consideration are water management (supply and storage, drainage and ditching, infrastructure), emergency planning and land-use practices.

This planning approach supports the recommendations made by ACT in its 2013 report *Climate Change Adaptation and Canada's Crops and Food Supply*, because it brings local governments, agricultural producers and government agencies together with a focus on implementation. The program provides up to $300,000 in each region for projects to implement actions from the regional adaptation strategy, so by the end of the five-year agreement there should be real changes on the ground.

The integrated farm-water planning project in the Cowichan Valley is a revealing example of interconnections across the nexus. A holistic approach to water management – supply, irrigation and drainage – is being applied at the farm level with the goal of developing a toolkit that has been pilot tested with ten farm operators. The toolkit can also be applied to more complex water–agriculture interactions involving multiple farms.

Drainage is vital for farm productivity in the Lower Mainland area of British Columbia, due to wet winters. In the Fraser Valley, factors such as increased flooding due to rapid runoff from impervious surfaces in urban areas, the potential for flooding from high freshet flows in the Fraser River and flooding due to sea level rise and storm-surge events combine to create a unique set of flood risks that must be managed in an integrated manner. The holistic approach being taken to these risks in the regional strategies for Delta and the Fraser Valley meets the test of true adaptation strategies. If the Ministry of Agriculture's climate change adaptation program is extended for a further five years in 2018, more comprehensive and longer-term approaches will begin to shift implementation to more transformative strategies.

Energy

In 2010 ACT released a report called *Climate Change Adaptation and the Low Carbon Economy in BC*, which focused on four policy areas: governance, energy conservation, adaptation and insurance. The report recommended widespread emulation of British Columbia's carbon tax; energy pricing that is reflective of externalities; and carbon offsets that support ecosystem restoration.

It is worth noting that in 2008 the British Columbia government introduced a carbon offset policy. All public sector institutions in the province – government buildings, hospitals, schools and universities – are required to be carbon neutral in their energy use. To the extent they still emit carbon, they have to pay $25 per tonne of emitted carbon into a fund called the Pacific Carbon Trust. The fund is then used to purchase equivalent offsets from businesses which have reduced their carbon footprint to less than that required by law and are accredited by an independent agency. The fund – over $37-million – was transferred to the Ministry of Environment in 2013 as a cost-cutting measure. The Auditor General heavily criticized the offset policy because much of the funding for corporate businesses that provided carbon credits amounted to a handout. In addition, there was concern that payments for net carbon emissions were taking money from front-line services in schools and hospitals.

British Columbia's electrical energy largely comes from renewable sources – hydro provides over 93 per cent of the total. Climate-related changes in hydrologic cycles will have a range of effects on hydropower projects. In northern latitudes and at altitude in the mountain ranges of British Columbia's interior, snowfall will continue and reservoirs will, in the short term, be supplemented by melting glaciers in summer, both leading to temporary increases in power potential. But in the longer term, these glaciers are likely to disappear almost entirely, with profound impacts on power production and water security.

In southern latitudes, however, rain will increasingly replace snow in winter, resulting in lower summer storage, which relies on spring snowmelt.

This is exactly what has already happened in California, where there has been not only less winter precipitation but most has fallen as rain and already passed through the reservoirs by May. Consequently, in the summer, with less water stored and increasing demands for power (think air conditioning, irrigation and in-stream flows for fish), water security will become the most critical element in drying climates such as the western states.

Coal-fired energy plants will not fare any better under a changing hydrology. Many such plants are located in areas where surface water will become scarce and groundwater resources are being depleted, such as the southern states, the Canadian prairies and China. Coal-fired plants require a lot of water for steam generation, and they have large carbon footprints. Ontario's long-term energy plan calls for an end to coal-fired generation by 2025, planning to replace coal-derived energy with a mix of 42 per cent nuclear, 46 per cent renewables and 12 per cent natural gas.

However, there are similar concerns about the consumption of water for natural gas fracking in Canada as well as related reduction in water quality for both surface and groundwater resources. British Columbia is in the process of preparing a cumulative impact assessment in its northeast region, where most of its natural gas fracking is occurring. The province's *Northeast Water Strategy*, released in March 2015, provides the first comprehensive look at all aspects of water use in energy production, agriculture and resource industries. First Nations in this region possess a deep affinity to water and a strong desire for water stewardship. The report is based on three key principles: water is not only a resource but a life source; First Nations traditional activities depend on water; and First Nations are committed to shared stewardship of the water resource. Based on these principles, most of the First Nations and the province jointly agreed to a unified approach to water stewardship across all levels of government. The primary focus of joint stewardship was to maintain healthy ecosystems, wetlands and ecological carrying capacity and protect riparian habitat and wildlife species. The strategy has already had an impact on the fracking industry, with the province placing a moratorium on water withdrawals from smaller streams, as a result of the drought in the summer of 2014. In September 2015 the Environmental Appeal Board, an independent administrative tribunal that reviews environmental approvals, cancelled a water licence application by

Nexen Energy for fracking on the basis that it would cause significant damage to fish and wildlife and that the First Nation affected – the Fort Nelson Indian Band – had not been adequately consulted and compensated.

Canada has adopted a sector-by-sector regulatory approach to energy adaptation policy. In 2012 the federal government enacted regulations identical with those in the US to increase fuel efficiency for motor vehicles. By 2025 new cars are to consume 50 per cent less fuel and emit 50 per cent less GHGs than 2008 models. The Ontario government is supporting this move by investing $235-million in General Motors for improvements in fuel efficiency and fuel-cell technologies in domestically produced vehicles. Canada has also adopted regulations that essentially ban coal-fired plants unless they enable carbon sequestration. However, it has not yet enacted long-promised regulations on the country's fastest-growing source of carbon emissions – the oil and gas sector.

ACT's *Climate Adaptation and the Low Carbon Economy* report also focuses on shifting to distributed energy systems in place of large-scale centralized facilities with long-distance transmission lines, which are becoming increasingly vulnerable to extreme weather events like ice storms, windstorms and wildfires. Ontario has introduced a system of "feed in" tariffs whereby businesses and homeowners are given subsidies to install solar power and feed any excess they generate back to the grid, thus reducing the cost of providing electricity. The program has been criticized for increasing electricity rates and for "windfalls" for owners who would have installed home-based energy systems anyway.

It is instructive to consider how a proliferation of decentralized solar panels on homes can force changes to the centralized distribution model. In Hawaii, for example, rooftop systems now sit on roughly 12 per cent of homes, by far the highest proportion in the United States. Making electricity at home, however, puts pressure on old infrastructure and cuts into traditional utility revenues. In sun-rich regions of North America, grids designed to carry power in one direction must now carry power both into and out of homes, causing voltage fluctuations and overloaded circuits, burned lines and even blackouts. Utilities in Hawaii are reacting by charging hookup fees or reducing the amount paid to customers for the electricity they send back to the grid. With the push to reduce carbon through national commitments, power utilities will have to further adjust their business model and their technology.

One answer to the problem may lie in avoiding the utility and grid completely by installing home storage batteries. In May 2015 Tesla Corporation announced that it was testing such home batteries with a view to building a factory in Nevada to begin mass-producing them in 2017. Speaking with Diane Cardwell of *The New York Times*, James Whitcomb, chief executive of Haleakala Solar in Hawaii, noted that "the lumbering big utilities that are so used to taking three months to study this and then six months to do that – what they do not understand is that things are moving at the speed of business. Like with digital photography [the shift to decentralized energy systems] is inevitable."

People and places are thus changing the game and implementing more radical adaptive approaches. Once people become aware of the changing climate and its potential impacts on the nexus, they start to demand changes to existing ways of doing business. As we have seen in energy, agriculture and water use in California, necessity is the mother of invention – there is a growing desire to try riskier approaches.

Another radical suggestion for the energy sector in Canada was proposed by a group of Canadian scholars in a 2015 report called *Acting on Climate Change*. They noted that 77 per cent of all electricity produced in Canada is from renewable or low-carbon sources – mainly from hydro and nuclear plants, but also solar and wind systems. They propose that a grid of high-voltage transmission lines between adjacent provinces could lead to a low-carbon future for all electricity production across the nation. Ontario and Quebec already have such a co-operative agreement. With British Columbia now committed to building the Site C hydroelectric project on the Peace River, Alberta could use some of the clean energy from the dam instead of the dirtier kind produced by coal-fired plants. Manitoba could link its hydro with Saskatchewan's grid. The authors of the report suggest that a low-carbon energy system could be in place across Canada by 2035.

THE COLUMBIA RIVER BASIN AND TREATY

Draining an area of 670,800 square kilometres, the Columbia River is the fourth-largest river in North America. Although only 15 per cent of this drainage lies in Canada, it produces almost 50 per cent of the flood flows downstream into the US. Although the United States constructed some 30 dams in its portion of the Columbia during the Great Depression and in the 1940s, it could still not contain the large volumes of water that originate in the Canadian part of the basin. In 1948 the Columbia flooded, causing massive damage to Vanport, Oregon. This event was the genesis of the Columbia River Treaty, ratified in 1964 between Canada and the US. The Treaty requires Canada to store 15.5 million acre feet of water behind three dams in the Canadian headwaters. (An acre foot is equivalent to 1 foot of water over an acre of land.) These storages together with the reservoirs in the US provide security from flooding that is estimated to be worth more than US$32-billion over the 50 years of the Treaty. The regulated flows crossing the border also immediately increased power production in the US, which was sold to Canada for $254.4-million over the 30 years between 1964 and 1994, after which time the United States paid Canada on average $150-million annually. The US storages collectively provide 29 gigawatts of capacity, or about 44 per cent of all the hydroelectricity generated in the US.

The Columbia River also provides fresh water for agriculture, cities and other communities throughout the basin. Roughly 2.4 million hectares are irrigated in the mid and lower reaches of the Columbia basin in the States, providing a good portion of

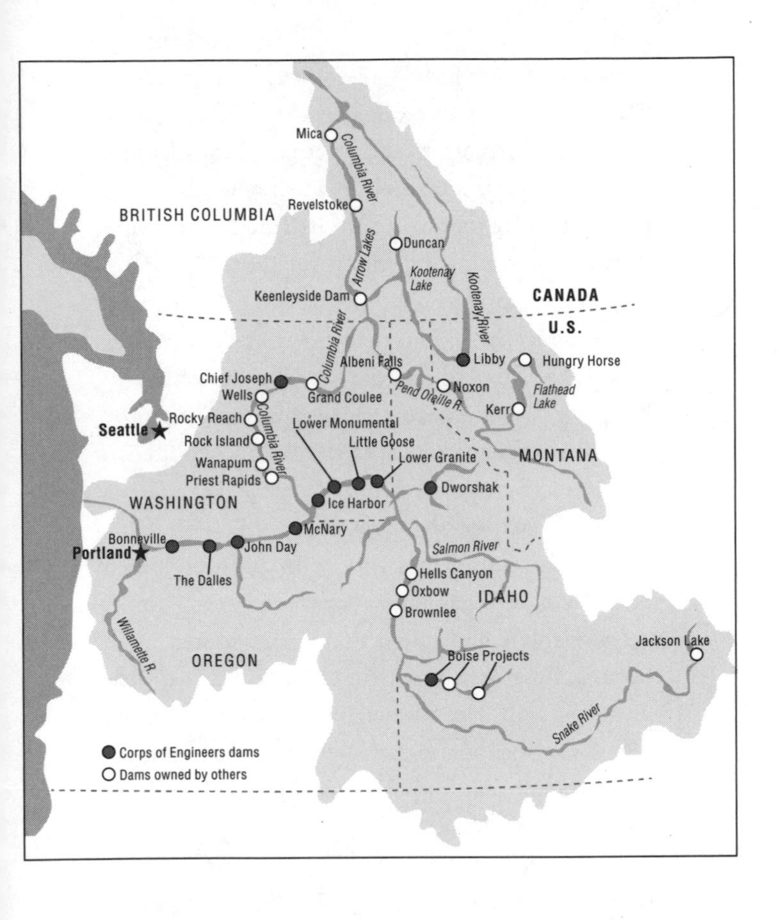

Figure 1. Storages in the Columbia River Basin

the over US$70-billion in agricultural values in Washington and Oregon. In times of drought, irrigators can pay up to $200 per acre foot of water. These withdrawals are projected to increase by 71–86 per cent by 2050 to support irrigated agriculture, thus heightening competition for increasingly limited water supplies. The water required to expand irrigation in the US may not exist, however. Garry Clarke and his colleagues from the University of British Columbia have warned that more than 90 per cent of the glacier ice in the Rocky Mountains and interior ranges of British Columbia that feeds the headwaters of the Columbia is likely to disappear by the end of the century. In addition, much of the winter precipitation which currently falls as snow in the American portion of the drainage will begin falling as rain as the climate warms. This means it will flow through the reservoirs more quickly in spring and early summer, leaving much of the water needs for power, fisheries and irrigation high and dry.

Even without climate change, the Columbia River Basin is a good example of the perfect storm of increasing demand for water across the nexus even as supply dwindles. Water security for all uses – including ecosystem health, food production, flood control and hydro – will become *the* strategic issue for basin residents before mid-century.

Under circumstances like these, it's easy to see why the Columbia River Treaty is crucial to the entire Pacific Northwest. And that is why the present authors, together with Deborah Harford of the Simon Fraser Adaptation to Climate Change Team, compiled a concise overview of the pact in their 2014 book *The Columbia River Treaty: A Primer*.

Essentially, the agreement deals with three storages in Canada: the Mica, Duncan and Keenleyside dams, plus the Libby storage in Montana, which floods into Canada. The pact does not directly cover the numerous other reservoirs depicted in Figure 1 on page 95, however. Neither does the Treaty include a legal termination date, though either party can give ten years' notice in or after 2014 if there is an intent to terminate or renegotiate terms and conditions. The existing flood-control conditions will change in 2024 regardless, as the United States prepaid Canada US$64.4-million in 1964 for 60 years of flood control. After 2024 the US will have to pay to use Canada's storages in the event of a flood threat.

The threat to the nexus in the Columbia basin is complicated by changes in US policy to restore salmon populations in the river and because of the changing hydro-climatic conditions across the basin over the coming decades. In 1993 the United States declared Columbia River salmon to be endangered under the federal Endangered Species Act, and over the past 20 years, the Bonneville Power Authority has invested almost US$13-billion in major salmon restoration projects. Salmon passage is blocked at the Grand Coulee Dam, however, so no salmon migrate into Canada up the main stem, though they now have access to the Okanagan and Similkameen river systems below the dam. Salmon require adequate river flows at critical times of the year for habitat – in fact, some of the Treaty-regulated waters from Canada are used for these flows rather than for generating power. In the dry summer of 2015, low flows and high water temperatures eliminated most of the returning salmon to the Okanagan system – a sign of things to come.

Climate change in the basin will have dramatic impacts on the nexus. Over the next 50 years, average temperatures are expected to rise by between 3 and 5°C. Stream flows in the lower Columbia could drop by 50 per cent, and most of the glaciers in the upper reaches are likely to disappear by the end of the century. The increased quantities of water vapour carried in atmospheric rivers will undoubtedly bring massive flooding potential to the Columbia on occasion over the coming warming decades.

ACT has emphasized the importance of maintaining healthy ecosystems in face of all this hydrological change, to increase resilience and provide valuable ecosystem services like water storage in wetlands, prevention of flooding, carbon storage in soils and improving water quality. The ACT report *Valuing Ecosystem Goods and Services in the Columbia River Basin* put it this way: "By supporting fisheries; regulating the quantity and quality of air, water and soil; attracting recreational visitors; and providing aesthetic, religious or educational value, ecosystems in the [basin] produce in total between US$19-billion and $701-billion of benefits to the regional economy every year." The value of fish alone is estimated to be between US$150-million and $600-million annually, based largely on current annual spending by US agencies and market values for commercially caught salmon in the Columbia. Recreation and the existence of nature itself are worth about another US$1.2-billion to $2.4-billion annually (all figures are in 2013 USD). As the report is careful to point out, "This large range in values represents an approximate appraisal of the natural capital in the US [basin] that replaces the former estimate of zero that has been the default value of ecosystems."

These values have never been included in accounting for the effects of climate change. But with both the US and Canada poised to begin their review of the Treaty after the 2014 starting date, some of these values will have to be brought into the negotiations. More traditional market-based values in agriculture are already available, such as a reduction in farm incomes by between US$48-million and $64-million in the Klamath River Basin due to drought in 2001. Overall impacts of reduced summer flows across the Columbia basin have been estimated to exceed US$1-billion annually.

Canada is in a commanding position in the coming Treaty negotiations, not only because it is upstream but also because most climate scientists agree that in the short run there will be increased runoff into Canada's storages, due in part to melting glaciers. The range of demands for water in the US is growing steadily, not only for the original purposes of hydropower and flood control but also for increased irrigated agriculture and for maintaining river levels for navigation in the lower reaches and water-based recreation throughout the basin. That is why regulated water flows from Canadian storages will become more and more valuable over time.

Ordinary residents in the basin, including representatives of First Nations on both sides of the border, have met numerous times to share ideas and ideals. People are aware of the coming changes in climate and hydrology, and they will be demanding a balanced approach to whole-systems management. Canadians, for example, are pushing for a change at the Arrow Lakes reservoir close to the US border, asserting that operating levels should be stabilized to provide a large area of riparian habitat for wildlife, forests and even agriculture. The proposed regime would mimic the historical fluctuation of the Arrow Lakes before the Treaty and go some distance toward reversing the ecological and social damage caused by the original flooding in the 1960s.

Individual residents are taking the lead in demanding changes needed to protect ecosystem values across the Columbia basin and not just under the constraints of the original Treaty. Some are converting their homes to use renewable energy, reducing demands from the centralized electricity grids; some are installing more efficient irrigation systems as the price of water rises; local conservation groups on both sides of the border are coordinating projects to restore fish and ecosystem health.

Co-operative regulation of Canada's storages under the Treaty will be critical to reduce tensions in the nexus in the Columbia. In the lead -up to the negotiations, two of the interest groups most alive to these issues are the Native American tribes and the Canadian First Nations. Among these peoples there is a unanimous chorus for better protection of ecosystems, which took a hammering in the original Treaty. The First Nations are articulate in their demands for restoring salmon into parts of the Columbia that have not seen the fish for over 70 years since the first US dams were built in the 1930s.

Reconsideration of the Columbia River Treaty is a unique nexus in its own right: a juncture not just between water, food, energy and climate but also a moment in time when history catches up with itself and reconciliation with the past becomes possible. Renegotiating the Columbia River Treaty could, in fact, become that moment in the history of both countries when ecosystem vitality is at last determined to be equivalent to and commensurate with social justice and equity.

FOUR

If We Fail: Truth and Consequences at the Nexus of Environment and Economy

It is common for individuals to bemoan the potential effects of climate change but at the same time carry on their daily routines, driving cars, heating their homes and consuming imported foods. Governments present policies that on the one hand commit to carbon reduction targets as outlined in chapter three even as they implement other policies that further embed increases in carbon emissions. Policy-makers in one government agency promote adaptive policies and procedures while another agency is promoting practices that contribute directly to the crisis in the nexus.

This dichotomy illustrates the "wicked problem" associated with climate change, reduced services from natural ecosystems and the inherent conflicts in managing water, food and energy in an escalating demand spiral. This is like the dilemma associated with driving along a long street with a series of traffic lights. Some are green and progress is made, but many are red and detours are required. We need all the traffic lights to be synchronized green; then we will make progress.

In a liberal democracy it is challenging for political leaders to run too far ahead of the voting public. Though there is a growing awareness that the climate is changing and that governments should set aspirational targets for carbon reduction, the public is generally not yet willing to incur additional costs to make adjustments. In Hawaii, for example, people want to adopt a target of 100 per cent renewable electricity by 2045, but this target is as much about the high costs of importing fossil fuels as it is about being a leader in renewable energy. We cannot move forward until we truly understand that the costs of inaction are greater than the cost of adapting to climate change together with the cost of reducing emissions. This is the message of this chapter.

THE COSTS OF INACTION

In his seminal work *The Economics of Climate Change*, British economist Sir Nicholas Stern concludes that the benefits resulting from strong early action to tackle climate change far outweigh the costs of not acting. Stern estimates that the costs of failing to address the impacts of a changing climate on the nexus, and the resultant loss of nature's ecosystem services, represent the equivalent of losing 5 per cent of world GDP per year into the future.

The rising costs of inaction are already becoming hard to ignore. The UN estimates that annual losses from natural disasters across the globe now total US$2.5-trillion. According to the Insurance Bureau of Canada, the costs of disaster relief related to storms affecting the nexus averaged $36-million annually in 1970s Canada; they now amount to over $1-billion a year. These numbers partly reflect the aging infrastructure across the country, but they also reveal greater frequency and intensity of flood events. In many cases, infrastructure is sound but not designed to handle these more intense storms.

The federal government is waking up to this crisis. In its 2014 report of planning priorities, Public Safety Canada stated that "the rising costs of natural disasters and the financial burden on the federal government is the country's biggest public safety risk." The former National Round Table on the Environment and Economy provided the first comprehensive, independent assessment of potential costs associated with the changing climate in 2011, using a climate model that assumes a 5.3°C increase in average temperature, the high end of the IPCC projection range. Estimates of costs that cover more than impacts on the nexus range from $5-billion per year by 2020 to around $40-billion annually by 2050. Longer-term projections depend on assumptions concerning potential temperature increases and carbon emissions, and range from $80-billion to $221-billion annually by 2075. But costs could be as high as $800-billion annually, assuming no significant change in public policy. Annual costs directly attributed to the changing climate, however, will become so significant that they will trigger a meaningful political response.

Primary agriculture across Canada accounts for some $22-billion – just over 1.2 per cent of Canada's 2014 GDP. Yet in 2001–2002, droughts in the prairie provinces caused $3.6-billion in lost production and resulted in over 40,000 people losing their jobs. The 2011 floods in Manitoba's Red River Basin caused $936-million in direct damages not including lost productivity. Lawsuits launched by First Nations arising from the flooding added another $1.15-billion to the bill. Heavy rains in British Columbia's

Lower Mainland in 2010–2011 caused $6.3-million in losses to vegetable producers.

There are now synergistic effects due to extreme events which have repercussions on both food production and supply chains. In late 2012 drought in the US Midwest resulted in the Mississippi being so low that barges could not take the corn and cotton crops to market; alternative transportation increased the price of those goods and also caused over US$7-billion in lost trade. The Army Corps of Engineers resorted to using explosives to deepen the riverbed.

In 2012 poor agricultural practices in the UK caused chemical runoff into rivers during extreme rainstorms. Soil erosion caused by loss of biodiversity and increased flooding due to rapid runoff from exposed hillsides cost the nation US$745-million, or almost 10 per cent of the year's total farm receipts. The changing global hydro-climatic regime is highlighting the consequences of wasteful practices in land use, drainage, infrastructure design and agricultural development. Extreme events that once occurred once or twice a century are now happening somewhere in the world almost every year, setting off a chain reaction of unintended consequences.

This new reality is challenging the traditional

methods for monitoring the economic health of nations. The common measure is GDP, which tracks the production of goods and services in national economies. GDP does not, however, measure or note of loss of biodiversity and associated ecosystem services. Further, in a perverse way, the cleanup following catastrophic floods or wildfires can add to GDP, as there is a burst of activity and investment in infrastructure and services, which *are* included in the national accounts. This method of monitoring national economies will, of course, have to be adapted as we deal continue dealing with the crisis in the nexus.

Another coming change to current business models is to subsidies that are willingly paid by cash-strapped governments worldwide to encourage investment in fossil fuels and underprice irrigation water, which encourages inefficient use (even in drought-stricken places) and protects poor agricultural practices in international trade negotiations. According to International Monetary Fund figures for 2015, if environmental and health impacts were included in the calculation, government subsidies to the energy sector alone were projected to be more than US$5.3-trillion. This figure includes an estimate of the costs of climate change and premature deaths from air pollution, as well as direct subsidies, plus other externalities such as traffic congestion and accidents due to low-cost fuels. The IMF estimates the subsidies for renewable energy to be around US$1.2-billion annually.

In short, despite the crisis we are experiencing, we still subsidize fossil fuels, there is no real price on carbon emissions and no direct value placed on protecting biodiversity.

STRANDED FOSSIL-FUEL ASSETS

Analysts such as Christophe McGlade and Paul Ekins have calculated how much of the current known reserves of fossil fuels will have to remain in the ground to prevent the likelihood of global temperatures exceeding an increase of 2°C. They estimate we could stay within this cap if we burned three times the amount of all known carbon reserves. More precise amounts vary depending on the particular fuel. For example, 80 per cent of proven coal reserves, 50 per cent of natural gas and 30 per cent of oil reserves would have to remain in the ground. This has led to climate activists encouraging major portfolio investors to begin selling off their fossil-fuel holdings. A number of large institutions such as Syracuse University – US$1.12-billion as at June 30,

2015 – and the Rockefeller Brothers Fund – US$827-million as at August 31, 2015 – have made such divestment moves in 2015. And in September 2015 the California legislature passed a bill that requires the state's two largest pension funds to divest their holdings in thermal coal.

Shifts in the market, particularly for oil and gas, are also changing the economics of access to resources. As a result of new technologies uncorking cheaper shale oil in the US, and Saudi Arabia pumping record volumes of low-cost crude, exacerbating an oil glut in the market, at this writing the international price of oil has fallen by about half in less than a year. This has had profound impacts on investment in Canada's oil sands. As much as 1.2 million barrels per day of future production has been put on hold since the downturn, amounting to billions of dollars in investment. This is a sizeable chunk, as production had been around 3 million barrels a day. New technology in oil-shale extraction is undercutting older extraction methods in the oil sands. So Canadian companies are slimming down on production and supply-chain costs. With increased competition from the US, reduced production in Canada will likely delay kickoff of major transportation projects such as Enbridge's Northern Gateway pipeline to

Pacific tidewater and TransCanada's Energy East line to the opposite coast until well into the 2020s.

It will take time to reduce production costs in the oil sands. "You don't realize 20–30 per cent cost savings within a year like we're seeing right now for the cost of drilling wells in the US," said Chris Cox, a Calgary energy analyst quoted in *The Globe and Mail*. "... from a broader perspective, the oil sands are once again very marginal in terms of rate of return," he said. Some predict that production will flatline at about 3 million barrels per day by the end of this decade.

CARBON CAPTURE AND STORAGE

Canada has passed regulations that require all new coal-fired electricity plants, as well as largely retrofitted existing ones, to use carbon capture and storage (CCS) technology. CCS is a process for capturing CO_2 emissions and transporting them to long-term storage – usually deep geological formations – so they do not enter the atmosphere. CCS applied to a modern coal-fired power plant is said to have the potential to reduce CO_2 emissions by 80–90 per cent. The technology is quite expensive so far, however. It can increase overall costs of power production by up to 90 per cent, depending on the distance to underground storage sites. In 2005 the Coal Utilization Research Council (US) suggested that investment in research and development will make costs of carbon sequestration more competitive with old coal-fired plants by 2025.

We must note, moreover, that carbon capture has not yet been proven to work. Some of the stored carbon may find its way into the atmosphere over time due to a variety of factors including geological shifts, difficulties mapping deep formations and the volatile nature of CO_2. Despite this, Canada is a leading proponent of CCS, with Alberta investing $1.24-billion over 15 years to fund two large-scale pilot projects.

The Alberta Carbon Trunk Line will pipe CO_2 from a fertilizer plant and a bitumen refinery to a producing oilfield for injection to increase recovery. Costs are projected to be $1.2-billion over 15 years, including operating, with grants of $63-million and $495-million from the federal and Alberta governments respectively.

Shell Oil's $1.35-billion Quest project will likewise receive government grants: $120-million from Canada and $745-million from Alberta. The operation will capture 1.1 million tonnes of CO_2 annually from an oil sands upgrader near Fort Saskatchewan and pipe it 80 kilometres to be stored in a saline aquifer some 2.5 kilometres deep.

If successful, these two projects could reduce CO_2 emissions by 2.76 million tonnes – equivalent to taking half a million cars off the road.

Even if CCS proves effective, however, global implementation will be slow. It would not begin to have an impact before 2025, by which time a large chunk of the carbon budget will already have been emitted. Only time will reveal whether this technology will be effective or not – time we unfortunately do not have.

What's more, even if CCS does succeed and gets fully deployed, about 74 per cent of global oil and much of global coal would still have to stay in the ground in order to meet the UN target. Only about 24 per cent of natural gas reserves would be affected due to the lower carbon footprint of gas. CCS could also have an impact on Canada's oil sands development, in addition to the changing market share. CCS will further increase production costs, but without it, the carbon budget may become too high for the new national target set by Canada for the 2015 UN Climate Conference in Paris.

ECOSYSTEM GOODS AND SERVICES

Nature – lakes, forests, wetlands, soils – is capable of capturing and storing CO_2 without expensive technology. In fact, about 30 per cent of carbon released into the atmosphere is due to destruction of natural storage systems. To keep up with the growing demands for food, the UN estimates that 6 million hectares of new farmland will be needed every year. But 12 million hectares of farmland a year are lost to soil degradation. In other words, we wreck this and other precious ecosystems (forests and wetlands) as we go. One study conducted by Anthony Foucher and colleagues concluded that the intensification of agriculture over the past century increased the rate of soil erosion sixtyfold. Good husbandry can have the exact opposite effect. Carefully tended soil allotments in urban areas contain a third more organic carbon than agricultural soil and 25 per cent more carbon. These are some of the reasons why allotments can produce four to 11 times more food per hectare than large-scale industrial farms.

There have been a few attempts to value ecosystem goods and services. From 2007 to 2010 scientists undertook a study of the boreal forest, a massive ecosystem that covers 58 per cent of Canada's land surface and is largely undeveloped due to its relative remoteness. The boreal provides a vast source of carbon capture, water filtration, nutrient recycling, climate regulation, fish and wildlife habitat and subsistence resources for First Nations. The Canadian Boreal Initiative conducted a full-scale evaluation of both market and non-market values of this ecosystem: forestry, mining, recreation, hunting and fishing totalled an estimated $41.9-billion annually; non-market values like carbon storage, water filtration and more totalled far more: some $570-billion per year.

In light of these relative values, the Canadian Boreal Forest Agreement, covering 73 million hectares, was signed in 2010. The world's largest conservation initiative, it involves a number of major forest companies, conservation organizations and First Nations. The agreement aims to improve sustainable forestry practices, permanently protect 29 million hectares from logging, protect at-risk populations of woodland caribou and increase carbon storage and forest-product life cycles to reduce carbon emissions.

As noted in ACT's 2015 Columbia Basin study, ecosystem goods and services support fisheries; regulate water, air and soil quality; and provide aesthetic and recreational services. The value of the basin's natural goods and services is estimated at between $351-billion and $1.2-trillion a year. Admittedly these values are not subjected to the rigour of the economic marketplace, but they do indicate significant non-market values that should replace the default value of zero that conventional economic studies generally assign to ecosystems.

These studies point to two conclusions. First, the value of ecosystem goods and services needs to be incorporated in new tools for managing the nexus. We cannot possibly meet the growing demands for food and water, and limit carbon emissions, if we continue to erode nature's services. Second, we will require a form of governance that builds trust among resource industries, environmental groups and First Nations, interests that are frequently in conflict. The above examples demonstrate the kind of innovative policy-making required to transform the status quo into a new approach built on improved knowledge, better communications and revitalized relationships between institutions that have often been bedevilled by conflicting mandates and positions.

The Ultimate Nexus: The World We Want

Besides the myriad interactions between water, food and energy in a world with reduced natural capacity to support these resources, this book is also about a second nexus: the linkage between historical ways of managing these resources and a future that will require big changes in order for us to avoid the already unfolding crises in the first nexus. We conclude that the 2°C global warming target will be breached – certainly before 2050 and possibly as early as 2030 – despite the momentum generated this year by the agreements between the United States and China and the Pope's historic encyclical. There is just too much inertia locked into current fossil-based energy infrastructure to achieve the required reductions in GHG emissions. The result will be a permanent change to the nexus and a need for a transformational level of adaptation.

We will require a revolution in thinking and approach on the scale of other such transformative global cultural revolutions like the shift from the Stone Age to the Iron Age, the Industrial Revolution and more recently the information revolution and advent of digital technology. The revolution we are talking about here is built on two principles: living within nature's means and a comprehensive education and awareness initiative to build resilient communities across all levels of government. We need a new ethic on how we produce and consume food, water, energy and natural capital here and around the world.

LIVING WITHIN NATURE'S MEANS: RISK, RESILIENCE AND RESTORATION

During the 20th century, technological innovation created highly efficient, low-cost energy and water supplies which not only met rising demands for nexus resources but also reduced commodity prices. This growth came at the expense of natural capital and resulted in an explosion of carbon emissions that have raised global temperatures by 0.85°C over the past century – costs that are not captured in national accounts. Massive subsidies for nexus resource usage has further distorted the economic calculus and greatly magnified the risk to nexus resources, creating the crisis.

We have to change this approach to one of living within nature's limits. This is the central theme of Pope Francis's encyclical on climate change and the environment, a 180-page treatise released in June 2015. Austen Ivereigh, the pontiff's official biographer, told Stephanie Kirchgaessner of *The Guardian* that the encyclical "captures [the Pope's] deep disquiet about the direction of the modern world, the way technology and the myth of progress are leading us to commodify human beings and exploit nature."

We have to transform *risk* to *resiliency* – enabling ecological, social and economic systems to adapt to changes to the hydroclimate – and to *restoration* or rebuilding damaged ecosystems. This involves a transformative approach to managing the nexus – one that will *restore* natural ecosystems, *enhance* nature's capacity to store carbon through regenerated soils, forests, grasslands and wetlands (rather than rely on expensive and unproven CCS technologies), *respect* nature's hydrological cycle, *build resilience* into ecosystems so they are better able to withstand changes in hydrology and climate, and *conserve* water by means of a realistic pricing system. This transformative approach must apply to both land-based and ocean-based management of the nexus.

The shift from increasing risk to building resiliency means swapping the current linear economy – based on using resources once and then discarding them – to one based on full-cycle principles where there is no waste. Our new global economic model must transform waste into wealth.

EDUCATION AND AWARENESS: REGENERATING COMMUNITY IDENTITY

How can governments motivate average people to wean themselves off fossil fuels, to value and conserve every drop of water, to stop wasting food? Most people are not moved by abstract ideas, even though they may be aware of the changing climate. They are more likely to get motivated if they are working or competing in communities where neighbours and others they respect are installing energy or water conservation measures that are both cheap and convenient.

Stephen Sheppard at the University of British Columbia notes that most people are too busy day to day to deal with climate change, but when people are shown tangible and practical projects undertaken by neighbours, they tend to become engaged. A sense of community and not feeling isolated helps drive change. Social media can plays an integral part in motivation and even competition. As well, technologies like thermal imaging can help people "see" problems; smart meters *show* us that reduced energy consumption means reduced costs, and the attendant reduced carbon or water footprint is just the moral icing on the cake. Of course, living within nature's means and building resilient communities will make a difference to the use of nexus resources as well.

Let's begin with water use. Canadians are among the most profligate water wasters in the world, but we are waking up to the costs of our habits. We have, at enormous cost, overbuilt water infrastructure to support the wasteful norm. Now we find we cannot afford to maintain and replace overbuilt infrastructure that increases the risk of public health disasters like the one that occurred at Walkerton, Ontario, in 2000.

We are realizing that we waste huge amounts of energy treating and moving water to where it does not need to be used, and that this wasted energy is accelerating climate change, which in turn is starting to damage to the very infrastructure we can't afford to maintain anymore. In short, we have created a climate-change feedback loop that is costing us even as it compounds climate change effects. If we break this vicious cycle of wasting water and the energy it takes to move it, we can save a lot of money.

Industry has demonstrated that for every dollar we save in water conservation, we can save up to four dollars in the chemicals and electricity required to treat wasted water. But conservation will only get us partway there. We have to quietly, calmly and systematically redesign vulnerable elements of our built environment and water infrastructure around new hydrological realities. Our changing hydrology will demand that we reconsider what we build and where we build it. We also have to ensure that all construction is supported by – and supports – natural ecosystem function. Ecosystem restoration must include source water protection.

Take agriculture. The global farming sector produces about 4 billion tonnes of food a year. Yet waste across the food supply chain, from production, harvesting, storage, transportation, marketing and consumption, totals between 1.2 and 2 billion tonnes, or 30–50 per cent of total food production. In a world where water, energy and food face increasing challenges from both changing climate and loss of ecosystem services, such waste is unsustainable. Understanding the nature of this profligacy will assist in developing policies to combat it.

In developing countries, waste primarily occurs at the farmer-producer level due to inefficient harvesting and poor transportation infrastructure. Losses can range between 35 and 80 per cent. In developed countries, waste mainly occurs at the marketing and consumer level. Canadian households throw away 2.1 million tonnes of food a year – enough to fill Toronto's Rogers Centre three times over and amounting to $1,500 per year for each Canadian family. We buy more food than we need: up to 50 per cent of purchased food is discarded.

Not only is all this food wasted but all the precious water and energy used to produce it is also squandered. A single egg requires over 200 litres of water, a gallon of milk some 3500 litres, and a pound of hamburger a whopping 7200 litres of water. Journalist Stephen Leahy has calculated the total water footprint per capita based on the amount of freshwater used to produce all the goods and services we consume. It amounts to 8000 litres per day compared with the 300 litres we consume directly in drinking, washing and toilet use. Most consumers do not know how much water goes into food production, which underlines the need for universal information and awareness initiatives to encourage both a change of diet and a war on waste. During the two world wars, food scarcity resulted in a wholesale change in habits related to buying and producing food. The diminution of water supplies that is already apparent in California and elsewhere in the western United States will rekindle that sense that nothing can be wasted.

Another fundamental reason to embrace nature's principles is that food and energy prices are not as stable as they were in the 20th century, when growth in supplies often exceeded demand and resources were stockpiled. Today there is a much closer balance between global supplies of food, energy and water due to income inequity, loss of natural capital and reduced stocks, resulting in increasing price volatility for commodities. The loss of grain harvests during Russia's 2010 drought increased global commodity prices; the global recession in 2007–2008 resulted in food price increases that triggered protests and riots in 48 countries. We cannot afford to lose another 11 per cent of the world's remaining natural areas to the expansion of agricultural lands. Such losses have other consequences: for example,

42 per cent of cancer drugs are derived from natural sources.

MANAGING THE NEXUS

The global economy can use three methods to manage increasing demand for nexus resources: expand supplies, increase productivity and/or transition to low-carbon economies.

Expanding supplies is completely at odds with the principle of living within nature's means. Water supplies would have to be increased by 30 per cent over 20 years, affecting the natural environment in the case of new dams and further depletion of groundwater. Expansion would also impact ecosystem services – Earth needs more forests, not fewer. Tensions would emerge as additional land demands clash with ecosystem service requirements and production of biomass for energy.

More troubling relative to expanding supply to meet demand is the rapid increase in per capita demand for meat as diets change with rising incomes in developing countries. Livestock consumption is forecast to increase by 75 per cent by 2050, compared with 65 per cent in dairy products and 40 per cent in cereals. Unless new forms of restorative agriculture come into existence, expanded livestock production will add to a global increase in annual carbon emissions and place a strain on energy and water requirements.

Economists like those at McKinsey and Company, a multinational management consultancy specializing in advising large corporations, estimate that about 30 per cent of projected increase in demand in agricultural, water and energy systems can be met by improvements in productivity. About 40 per cent of global food production relies on irrigation, much of which depends on inefficient systems from poorly managed aquifers. Although drip and trickle irrigation systems – as opposed to flood irrigation – have high initial capital costs, they reduce water and fertilizer use by over 30 per cent, as they carry fertilizers directly to the root. Inefficient irrigation systems not only waste increasingly scarce water but actually reduce potential new supplies through contamination resulting from nutrient-laden runoff.

With just a half metre of soil standing between prosperity and desolation, improving soils is an urgent matter. As Courtney White, author of *The Age of Consequences*, observes, not taking care of soil is like running a factory without maintenance. We are presently at a point where depleted soil conditions are colliding with accelerated soil erosion, loss of farmland and climate disruption. What is needed is a new agricultural model, an agricultural revolution.

In the face of limits to increasing production and managing demand, attention will be focused on shifting to low-carbon economies. Community engagement, education and use of social media can play an important role in shifting attitudes and helping people adopt more efficient water and energy practices. Low-carbon pathways require transition to new technologies and distribution systems. Mention has already been made of new solar power technology and the potential for storage batteries, which could transform energy generation and distribution. Communities must come together to make change. Business models based on monopolistic utilities with high-cost, centralized distribution systems need to be changed to decentralized systems where homeowners generate their own power and are encouraged to share it with the grid – and not be penalized for that as is sometimes the case now. Electric battery technology and hydrogen fuel cells could transform the transportation sector in a decade with the right signals to entice entrepreneurs. The keys to increasing productivity and encouraging new technology are elimination of energy, water and food subsidies, removal of energy taxes and placing a global price on carbon. The collective impact of such policies would increase resource prices, which are required to spur the improvement in productivity in traditional technologies at the nexus and reduce reliance on expanding supplies.

A ROAD MAP FOR TRANSFORMATION IN THE NEXUS

There is a crisis in the nexus of water, food and energy, caused in part by the rising demands for these resources from a growing population and in part by loss of biodiversity, which buffers the nexus against the impacts of changing hydro-climatic conditions. The scientific panel that advised the UN in advance of the 2015 Climate Conference in Paris concluded that because of the huge scientific risks of a global temperature rise above 2°C, carbon reduction targets should be set for a 1.5°C temperature rise and carbon reduction strategies should begin immediately to provide a global buffer if we collectively overshoot targets.

However, the national carbon reduction targets announced to date fall well short of the kind of changes needed to protect the nexus, and there is no assurance that even these weaker targets will actually be met. One barrier to success is the unequal distribution of wealth between developed and developing countries, with developing countries looking to expand their economies on the existing model of subsidized fossil fuels, cheap water and expanded land development for agriculture. Developed countries are beginning to uncouple their economic engines from fossil-fuel consumption through efficiencies and a shift to renewable sources. One solution that has not been implemented is to establish a global fund of at least US$100-billion annually by 2020 to assist developing countries in the transition. Only $10-billion has been pledged to date, though both the US and China committed to contribute $3-billion each to the fund in September 2015. Both nations emphasized the importance of adaptation and building an international commitment to foster resilience and reduce vulnerability.

Though many people worldwide are aware of the crisis in the nexus, the costs of reduction in nexus services are not yet as well understood as the perceived costs required to gain traction for public support for the steps outlined in this section. This poses a governance dilemma, as it is likely that until the costs of climate change are fully and directly recognized, there will be a limit to the degree of transformative action undertaken in democratic governance structures.

PRIORITY ONE: EDUCATION AND COMMUNITY ENGAGEMENT

Our focus here is on how people and places are affected by the crisis in the nexus. With this in mind, the first priority is to undertake a comprehensive education and awareness program to help people and communities understand the magnitude of the risks facing them over the coming years, and prepare them to take individual steps toward water and energy conservation. At a minimum, average citizens need to know where their water comes from, how much they use and what they use it for. People need to understand that there is no guarantee that the amount of water available to people now will be there in the future, and water will become more precious with each passing year. This appreciation should be advanced in conjunction with an enhanced understanding of extreme events – floods, fires, droughts, wind storms, pest outbreaks – that will inevitably occur across the globe. Communities need to understand that such events are part of a pattern. People need to be encouraged to take the necessary steps to adapt to their changing circumstances.

Just as farmers in British Columbia are learning new adaptation tools to combat floods, redesign out-of-date drainage systems, control urban runoff and be prepared for droughts, others need to be given similar tools.

There needs to be a new water, energy and food ethic based on conservation, cessation of waste, and design with nature, an ethic that permeates how people use these resources, how they design their homes and gardens, and how they develop the land. People need to know where they are located in watersheds and how watersheds function, and they need to commit to sustaining the water resource to protect against flooding and droughts.

Californians are already going through this learning transition. Some richer communities feel they are entitled to use the same amount of water they have always used, so long as they are prepared to pay the higher prices set by the state. But state water reduction targets are not being met, so officials are now bringing in regulations and flow restrictions to reduce actual use. In addition, social media is being used to shame profligate users by exposing excess. This transition will play out in Canadian communities too as the climate changes and the new conservation ethic takes firm hold.

People need to think about how and what they eat. They should be aware of food waste and take concrete steps to stop it. They should be encouraged to buy local produce to reduce dependency on imported foods. They need to be aware of how much and what they require for a healthy diet.

A switch from a beef-based diet would go a long way toward reducing the crisis in the nexus. Livestock emit 15 per cent of global carbon; beef production consumes nine times as much water as cereals; intensive livestock operations are responsible for 30 per cent of total loss of biodiversity globally. Yet, as mentioned previously, beef production is forecast to increase 75 per cent by 2050. Unless agricultural practices associated with beef production drastically change in the direction of sustainability there will not be enough water or land in the world to meet projected demand. Though there is little public awareness of these facts, policy-makers have to give serious consideration to dietary changes if we are to avoid a serious crisis at the nexus.

Governments should also encourage transition to renewable energy in homes and businesses, make it easier for people to make investment decisions, and change utility regulations to support decentralized energy systems. As demonstrated in the expansion of solar power in Hawaii, electrical utilities need to encourage people to switch to renewable energy and provide incentives rather than barriers to ease the transition. Once this change occurs, entrepreneurs will design more efficient energy systems. The same approach applies to water conservation. People should be aware of the opportunities to reduce water use, and water purveyors need to design affordable systems that encourage conservation.

All these changes require new approaches in schools and post-secondary institutions. Basic courses on preparing for changes to the nexus should be designed with adequate field projects so that students receive first-hand experience in conservation design. Courses should start in primary school and gradually evolve so that all students have a basic education in how to personally adapt to a changing world.

Education and community engagement on how to adjust lifestyles to deal with the crisis in the nexus is the long game. It underpins the strategies outlined below.

FISCAL POLICIES

The most important fiscal policies will remove subsidies supporting high use of fossil fuels and irresponsible water consumption and will put a price on carbon. As has already been noted, the IMF asserts that if all the costs associated with fossil-fuel use are accounted for, the total subsidy from the public purse is US$5.3-trillion a year. This figure encompasses much more than direct subsidies to energy companies in the form of tax breaks and reduced cost of gasoline. It also includes the social cost of over a million premature deaths a year worldwide caused by air pollution from coal-burning power plants, particularly in China and India. It also includes the cost associated with the impacts of the changing climate and hydrology by not charging a carbon levy, assumed to be $42 per tonne of emissions. Other costs include road deaths from the larger number of cars on the road because of the lower cost of fuel, plus the social costs of traffic congestion. The total subsidy amount is a controversial figure, but in publishing it the IMF compresses into one single metric all the myriad social and economic costs of a world based on cheap fossil fuel.

Nicholas Stern and other prominent economists have suggested phasing out subsidies for both fossil fuels and water. It is extraordinarily difficult, however, to get governments to implement such policies, due in part to the lobbying power of potentially affected interests and the challenges faced by elected decision-makers in selling change to a public that is increasingly concerned about climate disruption but unwilling to pay higher costs for fuel and especially water, which has long been considered to be a "free" good. Further, phasing out subsidies has to be undertaken on a global scale and not simply by individual governments, as companies are global in reach and will simply move their investments elsewhere unless there is a universal approach.

When removing unnecessary subsidies, care must be taken that basic human needs for water, food and energy are met as set out in the UN Sustainable Development Goals. Some subsidies will have to continue in order to iron out inequities and protect the most vulnerable in society. The green funds to support transition to an adaptive society must take these needs into account. This should be one of the main policies discussed at the Paris conference in December 2015 and afterward.

The second fiscal plank is to place a price on carbon emissions so that their impacts are accounted for when weighing the benefits and costs of energy production. Canada has struck an Ecofiscal Commission to consider various models for pricing carbon, ranging from British Columbia's carbon tax to Ontario and Quebec's cap and trade model and Alberta's carbon-intensity approach. In its first report, titled *Smart, Practical, Possible*, the commission reviewed a number of price signals that have been used to shift consumption of some of the nexus resources. One example was a price on water use in Singapore, a country faced with a major water shortage due to surging demand and limited supplies. The government brought in a water pricing system to cover the cost of distribution and treatment, and reduced consumption by 9 per cent. Other fiscal policies under investigation include traffic congestion in London, where there is a congestion tax, and pricing systems in the UK for waste disposal that reduced total waste going to landfills by 40 per cent.

The Ecofiscal Commission's specific mandate is to consider changes to the tax system that will result in pricing pollution in a way that encourages conservation and wise use but does not raise overall taxes. The commission is also examining ways to piggyback on some of the innovation being undertaken by local governments across Canada and to develop mechanisms to allocate costs fairly across the income spectrum so that the poorer members of society do not pay more than the rich.

As is always the case with subsidies, carbon and

water pricing have to be considered on a global scale. This remains the biggest barrier to effective governance in dealing with the nexus. But small steps such as British Columbia's carbon tax can lead to bigger steps as other governments consider such innovations.

INFRASTRUCTURE

All developed and developing countries need to invest huge amounts in water, sewage, transportation and energy infrastructure. Getting such investments right in face of the crisis in the nexus will go a long way toward managing the crisis. Much existing infrastructure is either obsolete or not designed to stand up to the intensified range of floods, droughts, fires and pests anticipated under the changing hydro-climatic regime.

In keeping with the principles of design with nature and closing the loop on waste, there needs to be a revolution in the design of infrastructure in the 21st century. We need to retain natural features such as wetlands and offsets from flood plains, and preserve some land from development because of the increased frequency of flooding.

The Capital Region of Greater Victoria is taking a second look at its needs for redesigning its liquid waste infrastructure. Up until now the region has not had to treat its wastewater, as offshore currents have been considered strong enough to avoid impacts of pollution. Both federal and provincial governments have directed the region to install a minimum of secondary treatment – removal of solids – by 2020. The initial plan was standard: a large central treatment plant on the shores of Esquimalt, one of the municipalities, with limited opportunity for water reuse or converting biosolids (sludge captured from solids in treated wastewater) to renewable energy. The community denied the region zoning rights for the plant, however, which set in motion a whole new approach to wastewater design. Now the focus is on a number of smaller plants distributed across the region, with nature's design principles in mind. Waste will be minimized; treated wastewater can be reused in commercial and residential developments ad to irrigate golf courses. Energy will be generated from biosolids, and heat in treated wastewater can be used for space heating in buildings where feasible. This is the kind of whole-systems approach that will be required in the future to transform infrastructure based on linear design principles to one based on closed systems, where biodiversity is enhanced and all "waste" is considered a resource.

In May 2015 a major conference was held in Winnipeg on the economic implications of the threat posed to Manitoba's already beleaguered infrastructure by hydro-climatic change. Organized by the Partnership for the Manitoba Capital Region, the meeting was attended by 165 municipal, provincial and private-sector leaders. Participants said it was one of those all too infrequent days when each of the speakers had the right message for the right audience. Many felt the province crossed an invisible threshold into a new mindset that day. A combination of effective speakers took the audience through the consequences of not acting on hydro-climatic change, all of them demonstrating that we are not helpless to respond. You could see the lights go on in the audience. One observer noted that this was the first time in ten years of working on the Lake Winnipeg eutrophication problem that real progress was immediately apparent. The shift witnessed at this conference was evidence of the potential to change the way we think and act at the nexus of water, food, energy and climate on the Great Plains.

SCIENCE

Scientists still have much to learn about how climate will change and how hydrology will shift as a result. Among the most important areas of study are Earth's oceans, relatively unknown compared with land-based and atmospheric systems. Scientists advising the UN Climate Panels indicate that as much as 90 per cent of the additional heat generated from increased carbon concentrations in the atmosphere is captured in the oceans.

ACT, in most of its major sectoral climate change adaptation reports, recommends that indicators of hydro-climatic change need to be regularly monitored so that policy-makers are informed as changes occur and can make adjustments to adaptation policies. ACT has initiated analysis of monitoring such indicators in municipalities, even as it studies development of what those indicators should be and how they should be measured, and will continue to recommend focused monitoring to support innovative adaptation practices.

INNOVATION

Many policy-makers feel that innovation in energy systems, water conservation technologies and energy efficiency will be central to solutions. Entrepreneurs are already designing more-efficient electric cars and higher-efficiency solar panels and wind turbines. Some are even exploring nuclear fusion, a breakthrough which has eluded scientists for half a century. Policy-makers have to be flexible and ensure that such innovations are encouraged. Many governments are conservative risk takers – this is their training and their culture. If we wish to make real progress in tackling the crisis in the nexus, we must learn to take bigger risks.

People will continue to want to have all the conveniences of contemporary life but in a way that reduces impacts on the nexus. The most immediately attractive shifts in use of nexus resources are those that support current consumer lifestyles but in a more efficient manner, such a conversion from fossil fuels to electric motors or even hydrogen fuel cells. Whether or not we can maintain current levels of convenience and prosperity in the Anthropocene remains to be seen.

INSURANCE

Insurance is the business of managing risk. As is noted in chapter four, the costs of weather-related disasters have increased tenfold over the past few decades. Canada still does not have private insurance to cover overland flood flows. The taxpayer assumes the risk through senior government subsidies. A significant component of the road map to transforming the nexus will be a shift in insurance policy and practice whereby both private and public insurance will coordinate, spreading the responsibility for managing the increased risks associated with changes in hydro-climatic regimes across Canada.

GOVERNANCE

We have already commented on the dichotomy in government where one arm is promoting strong policies on emissions reduction and water conservation while another arm is encouraging fossil-fuel consumption and irresponsible water use. Governments have to be redesigned to focus on whole-systems thinking based on nature's own design. Nature is beautifully integrated; it never operates in silos. The new policies of adapting to the crisis in the nexus will have to be similarly holistic. This is easier said than done, of course, as governments have always been compartmentalized into discrete policy areas that could hardly be any less integrated. Again, a revolution in government structure must undertake the enormous challenges facing humanity.

This fact is being recognized, not just in Canada but globally. As this book went to press, the UN announced global sustainable development goals for the coming 15 years and is poised also to declare a new decade of action and co-operation or equivalent aimed at understanding the role water plays in achieving sustainable development for all. The main focus of these initiatives is to find integrated ways for humanity to come to terms with how to live energetically but sustainably at the nexus of the Earth's available water supplies and arable land without further altering the composition of the atmosphere.

In the end, the entire human population on Earth is one. We are not simply tribes and nations that operate in competing spheres. If we are to solve the crisis in the nexus, we will have to act in concert as one overall system, and learn to co-operate and support each other in ways that we have never thought of before.

For better or for worse, we are all in this together.

POSTSCRIPT

The enormity of the transformation required to manage the crisis in the nexus simply cannot be canvassed adequately in one book. In addition, significant policy shifts will be announced by political leaders on both carbon reduction and adaptation in the nexus resources in the aftermath of the UN Climate Conference in Paris and the updated UN Sustainable Development Goals.

The authors plan to undertake ongoing research on the range of adaptation policies briefly outlined in the road map in chapter five, as well as to track all significant new policy commitments. A comprehensive analysis will be forthcoming in future books in the *Manifesto* series, along with the authors' reflections regarding the newly elected (Liberal) federal government, which has pledged to work with the provinces to strengthen Canada's commitment to carbon emissions reduction and to building green infrastructure that respects a changing climate.

BOOKSHELF

Axelson, Jodi, David Sauchyn and Jonathan Barichivich. "New Reconstructions of Streamflow Variability in the South Saskatchewan River Basin from a Network of Tree-Ring Chronologies, Alberta, Canada." *Water Resources Research* 45, no. 9 (September 2009). Accessed September 1, 2015 (full-text pdf), http://onlinelibrary.wiley.com/doi/10.1029/2008WR007639/pdf.

Berhe, Asmeret, Jennifer Harden, Margaret Torn, Markus Kleber, Sarah Burton and John Harte. "Persistence of Soil Organic Matter in Eroding versus Depositional Landform Positions." *Journal of Geophysical Research* 117, no. G2 (June 2012): G02019. Accessed September 1, 2015 (full-text HTML), DOI: 10.1029/2011JG001790.

British Columbia. *Northeast Water Strategy*. March 20, 2015. Accessed September 1, 2015 (pdf), www2.gov.bc.ca/assets/gov/environment/air-land-water/water/northeast-water-strategy/2015-northeast-water-strategy.pdf.

Canada's Ecofiscal Commission. *Smart, Practical, Possible: Canadian Options for Greater Economic and Environmental Prosperity*. November 2014. Accessed September 2015, http://ecofiscal.ca/wp-content/uploads/2014/11/Ecofiscal-Report-November-2014.pdf.

Canadian Boreal Forest Initiative. *Canadian Boreal Forest Conservation Initiative Framework*. Ottawa: Boreal Leadership Council, 2012.

Cardwell, Diane. "Solar Power Battle Puts Hawaii at Forefront of Worldwide Changes." *The New York Times*, April 18, 2015. Accessed September 1, 2015, www.nytimes.com/2015/04/19/business/energy-environment/solar-power-battle-puts-hawaii-at-forefront-of-worldwide-changes.html.

Childs, Craig. *Apocalyptic Planet: Field Guide to the Future of the Earth*. New York: Pantheon, 2012.

Clarke, Garry K.C., Alexander H. Jarosch, Faron S. Anslow, Valentina Radić and Brian Menounos. "Projected Deglaciation of Western Canada in the Twenty-First Century." *Nature Geoscience* 8 (April 2015): 372–377. Accessed September 1, 2015 (full text pdf), www.eos.ubc.ca/~mjelline/212_website/212_live/resources/Required-Reading/Clarke_2015_Projected-deglaciation-of-western-Canada-in-the-twenty-first-century.pdf.

Climate Action Tracker Consortium. "Climate Action Tracker." Accessed September 1, 2015, www.climateactiontracker.org.

Cotter, Anthony, and Sukhraj Sihota. *Valuing Ecosystem Goods and Services in the Columbia River Basin*. Vancouver: Adaptation to Climate Change Team, School of Public Policy, Simon Fraser University, September 2015.

Ehrlich, Paul R., and Michael Charles Tobias. *Hope on Earth: A Conversation*. Chicago: University of Chicago Press, 2014.

Follet, R.F., J.M. Kimble and R. Lal. *The Potential of U.S. Grazing Lands to Sequester Carbon and Mitigate the Greenhouse Effect*. New York: Lewis Publishers, 2001.

Foucher, Anthony, Sébastien Salvador-Blanes, Olivier Evrard, Anaëlle Simonneau, Emmanuel Chapron, Thierry Courp, Olivier Cerdan, Irène Lefèvre, Hans Adriaensen, François Lecompte and Marc Desmet. "Increase in Soil Erosion after Agricultural Intensification: Evidence from a Lowland Basin in France." *Anthropocene* 7 (September 2014): 30–41. Accessed September 1, 2015 (pdf), http://geosciences.univ-tours.fr/images/media/20150220114813-anthropocene_60.main.pdf.

Harte, John, and Mary Ellen Harte. *Cool the Earth, Save the Economy: Solving Global Warming is Easy*. N.p. [Calif.?]: Self-published, 2008. Accessed September 1, 2015 (full-text pdf), www.cooltheearth.us/download.php.

Intergovernmental Panel on Climate Change. *Climate Change 2013: The Physical Science Basis*. Geneva: IPCC Secretariat, 2013. Accessed September 2015 (full-text pdf), www.ipcc.ch/report/ar5/wg1.

———. *Climate Change 2014: Impacts, Adaptation, and Vulnerability*. Geneva: IPCC Secretariat, 2014. Accessed September 2015 (full-text pdf), www.ipcc.ch/report/ar5/wg2.

———. *Climate Change 2014: Mitigation of Climate Change*. Geneva: IPCC Secretariat, 2014. Accessed September 2015 (full-text pdf), www.ipcc.ch/report/ar5/wg3.

———. *Climate Change 2014: Synthesis Report*. Geneva: IPCC Secretariat, 2014. Accessed September 2015 (full-text pdf), www.ipcc.ch/report/ar5/syr.

Isabella, Jude. *Salmon: A Scientific Memoir*. Calgary: Rocky Mountain Books, 2014.

Kirchgaessner, Stephanie. "Pope's Climate Change Encyclical Tells Rich Nations: Pay Your Debt to the Poor." *The Guardian*, June 18, 2015. Accessed September 2015, www.theguardian.

com/world/2015/jun/18/popes-climate-change-encyclical-calls-on-rich-nations-to-pay-social-debt.

Klein, Naomi. *This Changes Everything: Capitalism vs. The Climate*. Toronto: Penguin Random House, 2014.

Leahy, Stephen. *Your Water Footprint: The Shocking Facts about How Much Water We Use to Make Everyday Products*. Toronto: Firefly Books, 2014.

Lovelock, James. *A Rough Ride into the Future*. New York: Allen Lane, 2014.

Marshall, George. *Don't Even Think about It: Why Our Brains Are Wired to Ignore Climate Change*. New York: Bloomsbury, 2014.

McBean, Gordon, and Dan Henstra. *Climate Change Adaptation and Extreme Weather*. Vancouver: Adaptation to Climate Change Team, School of Public Policy, Simon Fraser University, 2009. Accessed May 25, 2014 (pdf), linked from http://act-adapt.org/extreme-weather.

McGlade, Christophe, and Paul Ekins. "The Geographical Distribution of Fossil Fuels Unused When Limiting Global Warming to 2°C." *Nature* 517 (January 8, 2015): 187–190. Abstract accessed September 1, 2015, www.nature.com/nature/journal/v517/n7533/full/nature14016.html.

National Round Table on the Environment and the Economy. *Paying the Price: The Economic Impacts of Climate Change for Canada*. Ottawa: National Round Table on the Environment and the Economy, 2011. Accessed September 1, 2015 (full-text pdf), www.fcm.ca/Documents/reports/PCP/paying_the_price_EN.pdf.

Ohlson, Kristin. *The Soil Will Save Us: How Scientists, Farmers, and Foodies are Healing the Soil to Save the Planet.* New York: Rodale, 2014.

O'Riordan, Jon. *Climate Change Adaptation and Biodiversity.* Vancouver: Adaptation to Climate Change Team, School of Public Policy, Simon Fraser University, 2008. Accessed May 25, 2014 (pdf), linked from http://act-adapt.org/biodiversity.

O'Riordan, Jon, Erik Karlsen and Bob Sandford. *Climate Change Adaptation and Canada's Crops and Food Supply: Background Report.* Vancouver: Adaptation to Climate Change Team, School of Public Policy, Simon Fraser University, 2013. Accessed May 25, 2014 (pdf), linked from http://act-adapt.org/food-supply.

Pokhrel, Yadu N., Naota Hanasaki, Pat J.-F. Yeh, Tomohito J. Yamada, Shinjiro Kanae and Taikan Oki. "Model Estimates of Sea-Level Change due to Anthropogenic Impacts on Terrestrial Water Storage." *Nature Geoscience* 5, no. 6 (June 2012): 389–392. Abstract accessed September 1, 2015, www.nature.com/ngeo/journal/v5/n6/full/ngeo1476.html.

Pope Francis. *Laudato Si'*, Encyclical Letter on Care for Our Common Home. Vatican City: Libreria Editrice Vaticana, May 2015. Accessed September 1, 2015 (full text HTML), http://w2.vatican.va/content/francesco/en/encyclicals/documents/papa-francesco_20150524_enciclica-laudato-si.html.

Potvin, Catherine, et al. *Acting on Climate Change: Solutions from Canadian Scholars.* Montreal: UNESCO–McGill University, March 2015. Accessed September 1, 2015 (full-text pdf), http://biology.mcgill.ca/unesco/EN_Fullreport.pdf.

Rand, Tom. *Waking the Frog: Solutions for Our Climate Change Paralysis.* Toronto: ECW Press, 2014.

Sampson, Bruce, Linsay Martens and Jeff Carr. *Climate Change Adaptation and the Low Carbon Economy in BC*. Vancouver: Adaptation to Climate Change Team, School of Public Policy, Simon Fraser University, 2010. Accessed May 25, 2014 (pdf), linked from http://act-adapt.org/low-carbon-economy.

Sandford, Robert William. *Climate Change Adaptation and Water Governance*. Vancouver: Adaptation to Climate Change Team, School of Public Policy, Simon Fraser University, 2011. Accessed May 25, 2014 (pdf), linked from http://act-adapt.org/water-security.

———. *Cold Matters: The State and Fate of Canada's Fresh Water*. Calgary: Rocky Mountain Books, 2012.

Sandford, Robert William, Deborah Harford and Jon O'Riordan. *The Columbia River Treaty: A Primer*. Calgary: Rocky Mountain Books, 2014.

Saul, John Ralston. *The Comeback*. Toronto: Penguin Viking, 2014.

Schuster-Wallace, Corinne J., Robert W. Sandford et al. *Water in the World We Want: Catalysing National Water-Related Sustainable Development*. Hamilton, Ont.: United Nations University Institute for Water, Environment and Health, 2015. Accessed September 1, 2015 (full-text pdf), http://inweh.unu.edu/wp-content/uploads/2015/02/Water-in-the-World-We-Want.pdf.

Sheppard, Stephen R.J. *Visualizing Climate Change: A Guide to Visual Communication of Climate Change and Developing Local Solutions*. New York: Routledge, 2012.

Slade, Giles. *American Exodus: Climate Change and the Coming Flight for Survival*. Gabriola Island, BC: New Society Press, 2015.

Stern, Nicholas. *Stern Review on the Economics of Climate Change*. London: HM Treasury, 2006. Accessed September 1, 2015 via archived link from http://webarchive.nationalarchives.gov.uk/+/http://www.hm-treasury.gov.uk/sternreview_index.htm.

United Nations Department of Economic and Social Affairs. *Transforming Our World: The 2030 Agenda for Sustainable Development*. September 18, 2015. Accessed October 1, 2015 (full-text HTML), https://sustainabledevelopment.un.org/post2015/transformingourworld.

Vörösmarty, Charles J., Páll A. Davídsson, Magdalena A.K. Muir, and Robert W. Sandford, eds. *Motivating Research on the Science Communications Front: Conveying the Nature and Impacts of Rapid Change in Ice-Dominated Earth Systems to Decision Makers and the Public*. Summary of Findings from an NSF-Funded International Workshop hosted by the World Bank in Washington, DC, November 12–14, 2014. Accessed September 1, 2015 (full-text pdf), http://geo-prose.com/pdfs/motivating_research_high.pdf.

White, Courtney. *The Age of Consequences: A Chronicle of Concern and Hope*. Berkeley, Calif.: Counterpoint, 2015.

Wilson, E.O. *A Window on Eternity: A Biologist's Walk through Gorongosa National Park*. New York: Simon & Schuster, 2014.

World Bank. *Turn Down the Heat: Confronting the New Climate Normal*. Edited by Sophie Adams et al. Washington, DC: World Bank Group, 2014. Accessed September 1, 2015 (full-text pdf), linked from http://documents.worldbank.org/curated/en/2014/11/20404287/turn-down-heat-confronting-new-climate-normal-vol-2-2-main-report.

OTHER TITLES IN THIS SERIES

THE WEEKENDER EFFECT

Hyperdevelopment in Mountain Towns

Robert William Sandford

ISBN 9781897522103

ETHICAL WATER

Learning To Value What Matters Most

Robert William Sandford & Merrell-Ann S. Phare

ISBN 9781926855707

FLOOD FORECAST

Climate Risk and Resiliency in Canada

Kerry Freek & Robert William Sandford

ISBN 9781771600040

THE COLUMBIA RIVER TREATY

A Primer

Robert William Sandford, Deborah Harford & Jon O'Riordan

ISBN 9781771600422

SAVING LAKE WINNIPEG

Robert William Sandford

ISBN 9781927330869

ON FRACKING

C. Alexia Lane

ISBN 9781927330807

AN ALTAR IN THE WILDERNESS

Kaleeg Hainsworth

ISBN 9781771600361

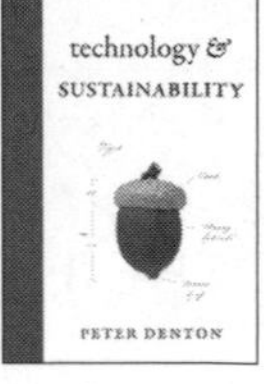

TECHNOLOGY AND SUSTAINABILITY

Peter Denton

ISBN 9781771600392

BECOMING WATER

Glaciers in a Warming World

Mike Demuth

ISBN 9781926855721

GIFT ECOLOGY

Reimagining a Sustainable World

Peter Denton

ISBN 9781927330401

DIGGING THE CITY

An Urban Agriculture Manifesto

Rhona McAdam

ISBN 9781927330210

DENYING THE SOURCE

The Crisis of First Nations Water Rights

Merrell-Ann S. Phare

ISBN 9781897522615

THE HOMEWARD WOLF

Kevin Van Tighem

ISBN 9781927330838

THE INSATIABLE BARK BEETLE

Dr. Reese Halter

ISBN 9781926855677

THE BEAVER MANIFESTO

Glynnis Hood

ISBN 9781926855585

THE INCOMPARABLE HONEYBEE

and the Economics of Pollination Revised & Updated

Dr. Reese Halter

ISBN 9781926855653

LITTLE BLACK LIES

Corporate and Political Spin in the Global War for Oil

Jeff Gailus

ISBN 9781926855684

THE EARTH MANIFESTO

Saving Nature with Engaged Ecology

David Tracey

ISBN 9781927330890

THE GRIZZLY MANIFESTO

In Defence of the Great Bear

Jeff Gailus

ISBN 9781897522837

RMB saved the following resources by printing the pages of this book on chlorine-free paper made with 100% post-consumer waste:

Trees · 9, fully grown
Water · 4,297 gallons
Energy · 4 million BTUs
Solid Waste · 288 pounds
Greenhouse Gases · 792 pounds

CALCULATIONS BASED ON RESEARCH BY ENVIRONMENTAL DEFENSE AND THE PAPER TASK FORCE. MANUFACTURED AT FRIESENS CORPORATION.